PRECESSIONAL TIME
AND THE
EVOLUTION
OF
CONSCIOUSNESS

"Using precession as a key, Richard Heath has harmonized the revelations of those great originals G. I. Gurdjieff, Alexander Thom, Ernest McClain, and the authors of *Hamlet's Mill*. This is the kind of synthesis one has long hoped for and a worthy sequel to Heath's own cosmological revelation, *The Matrix of Creation*."

JOSCELYN GODWIN, AUTHOR OF
ATLANTIS AND THE CYCLES OF TIME

"All mythology is rooted in the relationship between Earth, Moon, Sun, and Stars, and these stories are at the very core of human experience. In this book Richard Heath draws these ideas together to create a vivid and inspiring narrative that reconnects us with our true origins and future potential."

PAUL BROADHURST, COAUTHOR OF
THE SUN AND THE SERPENT

PRECESSIONAL TIME
AND THE
EVOLUTION
OF
CONSCIOUSNESS

How Stories Create the World

RICHARD HEATH

Inner Traditions
Rochester, Vermont • Toronto, Canada

Inner Traditions
One Park Street
Rochester, Vermont 05767
www.InnerTraditions.com

Text stock is SFI certified

Library of Congress Cataloging-in-Publication Data

Heath, Richard, 1952–
  Precessional time and the evolution of consciousness : how stories create the
world / Richard Heath.
      p. cm.
  Summary: "How stories enable us to identify the inner spiritual aspects within
our material world and participate in the evolution of human consciousness
foretold by ancient myths"—Provided by publisher.
  Includes bibliographical references (p.   ) and index.
  ISBN 978-1-59477-363-1
  1. Storytelling—Religious aspects. 2. Knowledge, Theory of (Religion)
3. Consciousness. 4. Evolution. 5. Cosmology. I. Title.
  BL628.7.H43 2011
  201'.3014—dc22

                                                                    2011012886

Printed and bound in the United States by Lake Book Manufacturing
The text stock is SFI certified. The Sustainable Forestry Initiative® program
promotes sustainable forest management.

10   9   8   7   6   5   4   3   2   1

Text design and layout by Virginia Scott Bowman
This book was typeset in Garamond Premier Pro and Gill Sans with Copperplate
as the display typeface

To send correspondence to the author of this book, mail a first-class letter to the
author c/o Inner Traditions • Bear & Company, One Park Street, Rochester, VT
05767, and we will forward the communication.

CONTENTS

ACKNOWLEDGMENTS

MANY SUBGENRES OF ANCIENT and modern mysteries as well as esoteric ideas have been integrated here without recognition of the background sources, perhaps to some readers' relief. Major influences have been Giorgio de Santillana and Hertha von Dechend, Mary Douglas, Ernest McClain, George Gurdjieff, Peter Ouspensky, and John Bennett. More direct influences have been Anthony Blake, megalithic pioneer Robin Heath, the metrologists John Neal and John Michell, and Carnac investigator Howard Crowhurst.

I am very grateful to Anthony Blake for reading, editing, and discussing much of this work as it was in progress. The megalithic itinerary of chapter 2 went through many revisions, with Robin Heath providing some editing as well as the all-important survey at Le Manio. Thanks to my wife, Jane, for supporting another book project.

INTRODUCTION

THERE IS EXTRAORDINARY POWER in the fact that it is only possible to do one thing at a time. This constraint not only reveals a primary characteristic of the universe we inhabit, but it also is relevant to how humanity composes its stories. Generally, narratives use a structure that moves a reader through *time,* because the parts of the narrative must be kept apart just as the objects involved are separate in *space.* However, a third aspect of narrative is that it cannot be known without *consciousness,* which is the means to know and to recognize the structure of things in the world of time and space. Quite simply, consciousness is the power to rebuild and put back together the world process, forming our narrative explanations of the world. Successive cultural groups each developed a new form consciousness that became embodied in their stories and monuments, creating the literature of our recorded history.

For example, the articulation of megalithic monuments can be seen as narratives, and, before these, the art found in prehistoric caves. The formation of our major language groups was also important in the sharing of stories about the world. In prehistory, the major subject of narrative was how the world came into existence. This involved gods modeled largely on astronomical phenomena. Many see these ancient tales as works of fiction and superstition, and the consciousness they developed through their narratives is dismissed or taken for granted. This dismissal is because consciousness is considered something that naturally belongs to human beings as part of our self-awareness. Furthermore, our picture of consciousness is that it is a constant human faculty, as if people in the past had a similar experience of it but had less understanding than us of the physical world.

HOW STORIES CREATE THE WORLD

The truth about consciousness is that humans do not own it, though they have the means to work with it. Consciousness is a cosmic energy that inheres to the universe as a fundamental characteristic: its *intelligibility.* This word, not much altered in our word *intelligence,* is the ability of systems to be understood. For systems to be intelligible, they must have a structure that not only functions as a machine, but also retains information about how and why it functions.

The ancients had systems of knowledge that purported to understand the how and the why of the whole world in grand yet simple schemes—schemes that implied that the world had been created according to a rather basic overall pattern and only rendered complex by the world process, through the repetition of these patterns on many different levels of scale. This is exactly how complexity arises in modern chaos theory, in which a simple algorithm can generate a very great complexity. Therefore the ancient cosmologies might seem conveniently simple while employing a master idea: that the creation of the universe might have involved a simple set of strategies or *structural principles.*

A universe revealing itself as intelligible would require a power not invented by the ancients themselves, but a power revealed to them through consciousness and the ability to see narrative structures within that universe. In other words, the ancients' works were of a developing relationship with consciousness, an energy that through us seeks to understand the world and its structures as part of a continuing creation in the world. The world represents to our awareness the *Other* relative to the *Self,* and it is this dualistic framework of relationship that generates the true narrative, which is more than a descriptive narrative, it being the life experience required for understanding.

Though ancient attempts to create a world narrative can be criticized as over specific or not based upon facts, consciousness was not always as concerned with facts as it is today, and indeed, facts are only part of consciousness. To order the world is the primary work of intelligent life, and most of that work, even today, deals not with facts but with *categories of experience*. It can be said that human beings are actually walking cosmologies in that they engage in and are engaged by *worldviews:* complex narrative structures that give the world a certain overall meaning. The scientific worldview is simply a member of a much wider set of possible worldviews, many of which are viable and useful as a context for the human activity of making meaning of the world. In the scientific narrative, the *Self* becomes an abstract observer and the *Other* a repeatable experiment within nature that reveals laws.

The question that arises is whether the power of narrative order, which is so ubiquitous, could be something beyond consciousness that, in the manner of the ancient cosmologies, created a world in which meaningful narratives would dominate human consciousness. In other words, if our intelligible life is made of narrative structures, could this be a property of the universe or, more specifically, a characteristic of our planetary environment as suited to the arising of intelligent life? The story of the arising of life certainly presents a narrative, recently revealed through biological science, and intelligent life appears to be like a cherry on top of that cake.

It becomes possible to conceive of a power beyond narrative that is the source of all the creativity that humans have put into their successive civilizations and cultures, and this is exactly what the ancient texts suggest—that there is a source of creativity in the heavens that operates over great time periods. This source can be called a creative energy, and its defining characteristic is its ability to generate completely new worldviews, whether great or small. When we look at successive civilizations, we are struck by the fact that they had unique worldviews that defined a whole program of activity within each civilization. It is also clear that the activities of a civilization can become exhausted and "wither on the vine" when its worldview ceases to be an overriding source of activity.

This creative power is symbolized in myth as a heroic character who steals fire from heaven, a story repeated with many variations. From myth, it is also clear that ancient sky watchers had observed the stars long enough to realize they slowly shifted their position in the sky relative to the Earth, a motion we now call the precession of the Equinox. Within this framework, the orientation of earth to the galaxy is also significantly changed, and from this came about early mythic speculations about changes on Earth *being caused* by these changes in the sky. These changes in the sky gave an apparent source for the narrative structure of time so that human narratives have an archetypal structure borrowed from the development of the universe itself. Prehistoric stories that have survived tell of proto-beings creating the world through speech, as if the universe was brought into being by storytellers.

In this respect, the causal powers of the universe shared with humans the powers of creativity and consciousness, and early human developers of this capacity became leaders in creating a bigger world of meaning. This assignation of divine powers is no conceit if the energies that are employed are indeed cosmic—that is, from beyond Earth, and not belonging to life itself, arising from Earth. Such a short circuit between the cosmic and human, in a sharing of common energies, would lead to the cosmic and human having the same narrative function in creative expression. To a culture that has now lost this direct sense of cooperating within a cosmic creative process, the messages from antiquity have become obscure.

We live in a culture that has, through vast resources and a simple strategy of factual causation, created a massive bubble of information, including remarkable reconstructions of history as far back as the prehistoric. As narrative, our own cultural development could not have happened without the developments in past epochs. The result is that we can look out over past time as if we are the custodians of the history of the human race. With this bubble of information, we are different from earlier peoples who did not have this information. Indeed, our identity and the challenges we can feel and respond to arise only due to the world we believe we inhabit, a world described largely by the information bubble in which we now live.

Thus we are at a moment of singular character, though we barely realize it. Can our bubble continue, or has it reached a peak before it bursts? What is the purpose of all this knowledge? Is it part of a completely new age dominated by information, or will it fall away to be replaced by something new? The view presented here is that a *new relationship to consciousness* is implicit in the information bubble within which we now live, and there is a new challenge for us within the expanded world created by it.

The area of the brain created by nature for imagining risk and opportunity is the prefrontal cerebral cortex. Here, an alternative mechanism has evolved for *remembering the "future"* that involves the generation of possible narratives rather than the mere collection and collation of what has actually been experienced. There is evidence that this function is largely below the level of conscious awareness for most people, and it seems likely that new powers to engage this function consciously are required to activate a new creative human activity: that of articulating the creative self.

What was formerly the purview of special, talented individuals or spiritual groups is becoming a necessary step for a wider humanity. To articulate one's creative self runs counter to the increasingly centralized control that has come to define the powerful role of materialistic states and globalized organizations that deny the soul of the individual except within the context of sanctioned types of activity. The growth of material knowledge has become a threat to humanity as a prison for consciousness, a golden cage. It cannot be defeated, but is rather the perfect condition for an evolutionary step for humanity. To become free of outside manipulation requires the activation of a new source of creativity—the creativity spoken of in the great precessional myths of the hero's journey to steal fire from heaven.

The seat of creativity is the divine spark, spoken of in innumerable ways over thousands of years. The sacred pattern placed within humans is *the source of something beyond existence* and coming from the origins of the universe itself. The primitive powers associated with the cosmic energies of consciousness and creativity can become conscious only by obeying what is true to the sacred pattern, through self-articulation, and this means learning to discriminate what belongs to our pattern, as a special contribution within the existing world, and our fulfillment thereby. In this, the human will is bound up in the act of making stories that then influence human activities in a profound way.

THE NOTION OF SPIRITUAL STRUCTURES

The failure to identify what is spiritual about the material world has forced those interested in spirituality to project a spiritual world into a place other than where we actually live. This confusion has been encouraged by the misreading of spiritual texts that employ literary devices within myth to imply the existence of spiritual worlds occupied by gods and other spiritual beings. These notions were in fact derived from the sky and its creation of objective time, which is cyclic and eternal, and made up of the ever-changing celestial relationships.

Myth was a subtle device to embody the natural structure found in this eternal world within a living tradition; it was never intended to be thought of as literally true. Like viewers meeting a soap opera star and talking to his or her screen persona, we modern humans have been trapped into creating religions that insist that such heroes become 24/7 aspects of reality, supermen operating from an invisible world to affect the personal lives of people. However, human existence can benefit from understanding the sky's elaborate drama.

This view of history has also given religious time a different form than was intended by the ancient storytellers, who instead were incorporating ideas of renewal and the cyclical nature of time into their stories. Modern time and history, even if imposed by God, tend toward the linear and involve progress along a line of time toward an absolute ideal, such as salvation. This paradigm has been reinforced by Darwinism, in which living species evolve toward remarkably ideal forms through a natural selection (survival) of the fittest. Both religious and scientific paradigms, therefore, have time as a line of progress—which has been entirely at odds with the main ideas of humanity up until the last two millennia.

In this book, we will trace how an older, cyclic view of time came to be displaced as human thinking evolved. Linear time does exist, but so does cyclic time, and beyond this lies another time that belongs to spiritual evolution. The notion of different types of time requires an unfamiliar cosmological vision, one which suggests that our present moment of here and now has, effectively, three rivers of time running through it: namely, time that passes; events that repeat; and steps that, once taken, create something new that is forever changed. Neither the ancient nor the modern views of time have incorporated the third type of time, because the time of real change, or *transformation,* is that within which the purpose of the Universe, a Creation, is being fulfilled.

The ideas of no god and of a god that interferes directly in the running of the universe both hide *the reason for a creator's absence.* The creators of the car engine do not have to intervene in that engine to make it work. Instead, once invented, the whole point is that the engine continues to be manufactured, improved, and used to produce motive power from oil products with a minimal intervention even from the driver and the trained mechanics of the modern car. Later, we will show how apparatuses (i.e., the entirety of means whereby a specific production is made existent or a task accomplished) such as the internal combustion engine evolve a higher energy (in the engine, of force) from a lower energy (of heat), through the excellence of its design and the *sacrifice* of the fuel's chemical energy, an energy higher than the mechanical force produced.

The combustion engine is a model for how the evolution of something higher from something lower can only come through the sacrifice of something higher still, which George Gurdjieff* called the Law of Three. When investigated, this law is found to be *universal* whenever something new is made. The victory of the engine is only a local gain, however, because energy in the material universe still has to be conserved. In the sphere of life on Earth, chlorophyll grabs light energy from the sun to grow and regenerate plants, yet the sun disperses vast energies without any such result. The sun is, therefore, the source of something higher that is sacrificed, in part, to feed life and provide the base of the food chain.

In this sense, the myths of prehistory and pagan times hit the nail on the head with their notions of a sacrificial god when used as a metaphor for the sun, in that the sun does not appear to interfere but simply keeps shining and rising every day. For modern humans, the sun is deemed a background fact within a curriculum that prefers, to the understanding of creativity, the visual spectacle and complex facts of astronomy. Yet facts, however spectacular, do not guarantee understanding, because creativity represents a higher energy than that of facts or measurements. The ancients approached understanding the phenomena facing them in life, with its cyclic framework of renewal and perishing, through similarly structured narratives.

Understanding, unknown to science as a higher energy, acts to evolve how a person takes in his or her world. In past epochs, understanding was considered a form of genuine initiation; today, we can still see this is true with regard to our technical specialists who effectively live in a different world than those not initiated into their mysteries. Yet beyond the specialists of any age lies a body of understanding about how the spiritual world evolves humanity itself, and that is what concerns us here.

The evolution of living forms created human beings who can think, giving them a massive evolutionary advantage in the natural world. The material resources of Earth have been unlocked, but the thinking aspect

---

*G. I. Gurdjieff is substantially referenced in chapters 5 and 6.

of humanity now appears at odds with both evolution (because natural selection often does not apply to humans today) and the environment (as species and ecology are in competition with material progress). Is this an unplanned consequence of evolution, or is there simply no plan? A possible explanation might be that this crisis is resolvable only through the evolution of humanity itself. In other words, the problem is an evolutionary test that involves a risk of failure, uncertainty, and hazard.

It is possible to see human evolution as having a framework that does not come from within humanity or even from within organic life. Instead, such a framework might be the continuing sacrifice of a higher energy so as to keep open the possibility of human evolution. Such a framework appears in myth when it alludes to the precession of the equinoxes as the skeleton of Great Time. Precession requires that our large moon orbit Earth to maintain the stable tilt of the poles, causing balanced seasons on the planet relative to the sun. The north and south poles describe a circle in the sky over thousands of years that forms an "arctic circle" in the polar regions of the solar system. This cycle of time came to be seen as divided by the twelve constellations—the star signs—one of which holds the sun, at the spring equinox, in different epochs: the Precession of the Equinox taking about 26,000 years to repeat in the same sign.

Such understanding of cosmology by ancient man might seem implausible, but there is increasing evidence that different revelations were introduced episodically during the current precessional cycle. These revelations manifested themselves through different breakthroughs in abilities demonstrated by humankind, starting with cave painting and continuing to the causal thinking (commonly called *reason*) typified by the Classical Greek culture. Further exploring the idea that understanding is a higher energy, we see that the successive breakthroughs in understanding take a narrative structure, because, without the foundation laid by previous epiphanies, what comes later would have been impossible to achieve. In our metaphoric sense, then, humanity's evolution is possible through the sacrifice provided by the invisible sun of each precessional age, each representing a new level of human development. This great cycle of time is more than that of fate and tragedy. It is the destiny of the fire stealers to inaugurate a new age, as revealed through the generic narrative or, as it is called by Joseph Campbell, the heroic *monomyth*. Indeed, in the mysteries, initiation focuses on such heroic narratives, which recapitulate the successive realizations of human potential. Each age has to find and steal a new type of fire before it can progress. Today, our fire is that of technology. Yet the path of technical knowledge is rapidly approaching its limits in developing human consciousness, partly because of its implicit materialism.

The rational materialist always looks for a cause—the place from which something comes. But the availability of higher worlds means that these worlds must exist seamlessly, side-by-side, with the world that can be quantified and yet not be quantifiable. *Synchronicity of events* is one candidate for the spiritual that cannot be explained, and another is found in the *properties of wholeness* within artificial and living structures. The moon is a causal conundrum: the likelihood of its existence being virtually zero and yet, being dead, it appears to have created the conditions for life on Earth.

If the ancients had a natural rapport with a spiritual world, then they would have had no need to develop a *synthetic* understanding of it; thus, it might have been the tragic loss of such a rapport that was a prelude to the development of modern rationality. The journey from being naturally aware of the world to understanding it would then manifest through a set of stages, making human history a developmental narrative.

The only cosmic cycle that could provide a framework for human evolution on the grand scale proposed here is the precessional cycle. The significance of the moon in both life and precession is confirmed by both modern science and ancient myth. If we created a higher dimensional device to maintain a constant but cyclic mechanism that could enable intelligent life to evolve to a state of higher intelligence, it would have to provide higher energies that were introduced over thousands of years according to the synchronicity of cosmic factors. If, additionally, we think about the narrative order required of a zodiacal sequence, we conclude that it must involve the introduction of new faculties into intelligent life, much like the introduction of characters, conflicts, and plot-driving scenarios into any story—those elements that make a story work.

Without a storyboard of general influence, there would have to be scriptwriters working for evolutionary history. The combustion engine would then have an angel instead of a carburetor. Precession provides a connection to the will that created the universe as a whole—its pattern and the human pattern. Providing a cosmic cycle *instead of a cosmic hierarchy* enables evolution to be self-organizing, as an expression of a universe that only uses spiritual powers such as synchronicity and that has a preference for a specific sacred ordering of things. Such an approach appears weak but it gives independence to the created world while providing a source of redemption from being an entirely determinate world. If the precessional framework sets up conditions for the synchronicity of events, with regard to likely outcomes, then it is within its gift to make things happen almost magically. What was impossible earlier can then become possible under changed cosmic conditions, but such conditions will be hard to causally identify, because they are lost within the mix of concurrent processes or encrypted in the configuration of the heavens.

All this may seem a recipe for superstition and fancy, but, like all hypotheses, a natural explanation is preferable (in the absence of any other credible explanation) for phenomena such as the creation of the complex language groups, the organization of megalithic complexes, the arising of complex musical theories, and the birth of causal reasoning. Without an explanation for these events, we cannot properly understand how such definite steps in human civilization occurred from nowhere.

In other words, it is possible to understand precession in terms of an evolution of the human mind and then to understand intellectual history as part of a coherent narrative. Once we have seen the pattern of the past, we can better know where the story might be traveling next. We can also see new human faculties that exploit brain functions already developed through environmental competences and their natural selection. For example, our ability to appreciate an imagined narrative comes not only from understanding stories in nature and society but also from our innate capacity to evaluate risks and opportunities in the future.

In fact, the natural environment is highly likely to develop brain functions that relate to the intelligible nature of the universe since understanding the environment is key to survival. In an intelligible universe, consciousness would have to be part of the power that organizes the universe prior to the understanding that humans can have of that universe. The structure of the natural environment could then be prepared for intelligent life through natural selection. Humans would then have evolved to tell stories like their gods, and we can then come to see the *intellectual environments* of successive societies as having developed new types of human intelligence.

Chapter 1

LANGUAGE AND THE CONSCIOUSNESS WITHIN STORIES

IN RECENT YEARS IT has become much clearer that the developmental stories, of life itself and then of modern humans, tell of catastrophic setbacks, each preceding a new stage of development. These catastrophes came in the form of ice ages, volcanoes, and impacts by small interplanetary bodies. If such catastrophic events had not occurred, it seems life would not have evolved into complex multicellular forms and hence never given rise to a human intelligence that can discover and contemplate this fact today.

THE STORY OF LIFE

Up until 600 million years ago, life was unicellular and was either dispersed throughout the sea or covered the surfaces of rocks. Single-cell life had been terraforming the earth, affecting its geology and especially its atmosphere. Levels of oxygen in the atmosphere had risen toward modern levels due to these single-cell organisms' photosynthetic conversion of carbon dioxide into oxygen. The stromatolites are a typical example of this process, in that they trapped mineral grains within a biofilm of cells, forming a layered accretion column between the shallow sea bed and the surface of the water, a biofilm that used light and carbon dioxide to build new cells, leaving oxygen as a by-product. This oxygen would later provide fuel for multicellular animals that could move by burning oxygen. And it took a catastrophe to create such animals.

This event was the greatest of ice ages, in which the entire surface of the earth was covered in snow and ice, so that the earth reflected the light and heat that normally warmed it. Something within Earth, its own volcanic mechanism, eventually, after millions of years, broke this climatic deadlock. Volcanoes emit large amounts of carbon dioxide and other greenhouse gases, and at this time, a large number of volcanoes erupted around Earth, warming it and causing the ice to retreat from the equator.

Following this severe ice age there was an explosion in the diversity of life, through the development of a new ability in those few cell types that had survived: the ability to connect themselves to each other, using the sticky protein called collagen. The first multicellular structures were a loose collaboration that gained some benefit by literally sticking together. Each cell still reproduced by cell division, so these multicellular structures could not yet reproduce themselves as a whole; they relied on structures in which individual cells naturally connected to other cells by using simple rules. These same rules lead to the fractal patterns still visible in ferns and sponges. The new genetic instructions required for these simple aggregates were therefore few in number but crucial, for without them the later development of multicellular animals would have had no starting point.

Another innovation from this time, seen in corals, is the release of eggs and sperm into the water to fertilize new life through sex between different individuals rather than through cell division or cloning. This initiated genetic evolution through the intermixing of genes between parents and the consequent possibility of genetic variety within which natural selection (of the fittest) first became possible.

However, the ultimate step for multicelled organisms was the creation of a universal template seen in trilobite fossils. A mouth was developed to eat, a midgut to digest, and a hindgut to excrete. A gut that could digest food became the first of many new organs that would evolve in the middle section, while legs would form symmetrically to right and left. Hunting and eating required sense organs at the front end, and all animals came to evolve from this singular template.

Without a "snowball earth" this development would never have occurred, and after billions of years of life on Earth, it took this event to cause the unprecedented explosion of new forms of multicellular life.

END OF ANOTHER ICE AGE

The story of modern human history only became possible after the end of the last ice age about 12,000 years ago. We know that humanity was tenacious in survival during that ice age, but human civilization requires conditions that mere survival cannot provide. Knowledge must be passed down through the generations and communicated between communities to develop any form of civilization. Communication and retention of important information comes down to the use of language, and the first ancestors of modern languages arose around this time, languages that were to be the root languages of the post-ice human race. Just as the trilobite articulated the first true multicellular animals so too the root languages would come to define what could be said with a new sophistication in its articulation.

A parallel can therefore be drawn between single-celled life that could not develop 600 million years ago and the small groups of people who could do little more than survive the last ice age, again illustrating how an ice age can initiate a new structural development. The single cells that survived the snowball earth rapidly went on to clump together into larger communities of cells and, as these arrangements evolved, the cells began to develop intercommunication and true multicellular identities. The new units of organic identity developed organs, specialized cells, and new senses, as well as sexually reproducing itself as a population while renewing its individual cellular makeup through the earlier mechanism of cell division.

After the last ice age, it was the new environmental conditions that enabled surviving human groups to find new ways of interacting based upon agriculture, which replaced the hunter-gatherers with the herder-cultivators of the Neolithic revolution. The ability to grow communities beyond small groups and to trade between fixed territories led to production surpluses—a prerequisite for cultural and technical creativity—while supporting new and further specialization and value constructs, such as how to live with your neighbor. Neolithic cultures were therefore like multicellular organisms relative to the Stone Age hunter-gatherer family groups that were more like the single cell organisms of 600 million years ago.

The Fertile Crescent, east of the Mediterranean Sea, is considered the home of the Neolithic revolution, from which Neolithic culture diffused. However, other areas of Neolithic development probably existed, but these later became deserts or simply were lost to the historical record—the Sahara, Gobi, and Takla Makan deserts of today as well as anomalous regions, such as the Arctic coast, could have been homelands for this new, "multicellular" way of living.

THE VALUES OF TOGETHERNESS

The fundamental lesson of being part of a larger whole is that the other parts of the whole have value for oneself. This has been found in *altruism,* the genetic behavior, otherwise inexplicable, that probably led to the first multicellular structures. In altruism, one may look after someone else, because the other's survival is somehow going to aid the survival of ones own genes. Genetic-level altruism is therefore an instinct rather than a conscious decision. But when human families started to cooperate, within an extensive settled tribe, their fate became entwined in ways other than those of immediate genetic transmission; for, while their survival became tied to the survival of the society, the organization of the whole also expanded the cultural potential based upon conscious developments.

It is highly likely that the values that emerge within larger groups cannot develop in isolated groups and that these values naturally converge on a contemplation of the connectedness of the human world with that of the natural world. It is knowledge of the latter that led to new tools and technologies. Also, the natural and human worlds evoke human values, and this response develops an imagination of what is *possible,* either as a future reality, a new strategy, or as explanations for natural phenomena.

In the existential domain, Neolithic peoples generated material surpluses that could build and support cities and civilization. But in such a coming together a new type of vision was created of the interrelatedness of both man and nature. This vision became a religious *value system* made possible through being part of a larger agricultural whole, based upon a cultural altruism that then saw the natural world and creation at large as being in relationship.[1] Its shadow was the war between communities based upon instincts of a more genetic characteristic, that of gaining advantage over others.

The effect of these new values was to set humans thinking in new ways that would not be possible without the larger sense of self found in the cultures of interdependent communities. The diversity of these larger human groups revealed the poverty of smaller groups in pursuit of cultural variety. Like single cells, smaller groups are all the same and they simply reproduce as before, while larger groups can differentiate a wide range of possible cell types, each performing different functions that are undreamed of when you are a clone of your parents.

Because of this transformation in values, the factual world has not fully explained what such cultures were made of, though it does define its prerequisites. It was therefore values, awakened by participation within a greater culture, that became the source of cultural innovation for Neolithic peoples over and above the facts of life: food, shelter, and reproduction.

The coming of such values within Neolithic cultures corresponds with the development of new languages used to express what those values revealed about human nature and the human environment. Language then provided the building blocks of longer narratives, which we recognize now as essential to civilization. Some Neolithic stories have survived through the oral tradition of that time, and one of their common themes was found to have a bearing on precessional time and the idea that consciousness came to the human world from the cosmic structures seen in the sky. The appearance of consciousness as a subject within stories may be no coincidence.

THE STRUCTURE OF STORIES

When value-laden stories emerged, they were dramatic so that, by the late Stone Age, the conscious awareness of selfhood and otherness, which lies at the heart of dramatic reality and the human predicament, was inherently woven into their narrative structures.

Between the beginning and the ending of a story, a drama occurs: a challenge and a response are created—a conflict such as man versus man, verses nature, versus evil, and so forth. For consciousness, the challenge comes from the environment—the *known*—and the response from the *knower* (the primary actor). In a story, there is often a single actor, a protagonist, and therefore a hero who becomes a proxy for the reader. The challenge, or conflict as mentioned above, is often portrayed as an opponent, an antagonist, or a difficult state of affairs.

The literary structure of a story can be represented by a triangle within a circle. (See figure 1.1.) This structure, or form, was found to exist, not always consciously, within ancient texts—see Mary Douglas's book, *Thinking in Circles*.[2] This circular literary structure is related to the way the universe works, not just for stories but perhaps for all cyclic processes, especially when it comes to understanding how these processes work and what they mean, a point we will develop throughout this book.

In figure 1.1, the topmost point of the triangle represents the nature of consciousness as the synthesis of knower and known. This topmost

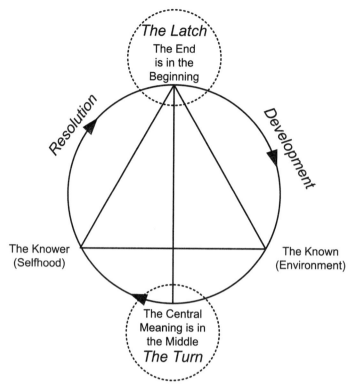

Figure 1.1. The form of the story cycle, structured according to consciousness of the knower and the known

point is both the beginning and the ending of the story. What occurs between is the timeline of the drama itself: the course of the knower interacting with the known and the resulting evolution of consciousness in the knower. The events portrayed in the story are projected in the mind's eye of the reader and are subliminal because of his or her own identification with the drama—as if it was actually occurring to him or her. This is the basic technology we recognize in the storyteller's art.

The base of the triangle in figure 1.1 marks the separation of fact and value,[3] given that the self is a reservoir of values and the (environmental) other is the reservoir of facts. The first half of a story is about generating the necessary separation between the self and the other, and the second half is about the resolution of this separation: the self becomes unified with or reconciled to the other.

- In this cyclic view, the beginning point is connected with the endpoint, which leads to a fulfillment, completion, or resolution of what was begun. The end completes the beginning, and this completion is said to form *the latch* between a beginning and an ending. The latch of a gate brings closure, and so the academic tradition of story analysis uses this term since it also expresses the moment at which the story as a whole finally clicks shut.

- There is also a turning point in the story cycle that occurs diametrically opposite to the latch. A crucial crisis of meaning arises at this middle point—called *the turn*—that expresses the primary meaning, often the greatest danger in the story. The turn gets its name from the change in direction, away from development and toward resolution, that occurs near the midpoint of the story. Mary Douglas adds that the central meaning of a story is often expressed at this central point.

The story form, or structure, therefore became a medium for transmitting a natural wisdom about beginnings, endings, and what happens in between. This structure directly parallels the way learning and achievement work for human beings in the world, in that there are stages of development and an overall aim.

- The right side of the circle in figure 1.1 represents the building of the stories' natural dramatic possibilities, and therefore it represents a *path of descent* into the challenges of the world it presents.

- The bottom of the circle in figure 1.1, the turn, is where the main work is achieved—a mythic smithy; a dragon's den; a princess imprisoned deep in the earth; and an escape with something essential, new, and valuable, such as a golden fleece or the philosopher's stone.

- The left side of the circle in figure 1.1 represents the place where what has been set up and developed in the first half of the story must come to culmination or climax, and the hero must be tested, tempted, and found fit. It is a *path of return*.

The overall scheme of the circle is that of renewal through an action that must necessarily engage with risks and the unknown, a hazardous or dicey business.

Narrative journeys are prototypes for history in that cultures always have their stories and their heroes, and these victories create a cultural cave within which the spirit of the time resides. The period in which languages were first used to create great stories made such a cultural cave, and their development was therefore a milestone in the history of consciousness. Language development can therefore be considered *the latch* in the story of human cultural development: a technology that engaged with values and enabled stories that would define the world each culture lived in.

LANGUAGES FOR TELLING GOOD STORIES

The earliest myths began a tradition of creation stories in which a unified state of being existed at a beginning of time, a Time Zero. Some of these creation myths appear to have used the precessional effect—of changes in the stellar environment—to explain that beginning as a past orientation of the sky within precessional time. If great stories only started with the Neolithic, then the new languages that developed around this time might represent a real beginning, or time zero, for the current precessional cycle.

The Indo-European language in particular, forming the basis of writing and speech in Western cultural and technical life, was one of the four or five root languages created about twelve thousand years ago. There is astronomical evidence in the earliest Indian and Persian texts that the Indo-European language first appeared in the northern limits of Europe and Asia, from where it spread. The evidence of a northern location for the root of the Indo-European languages comes from a popular subject for early myths: the behavior of the celestial objects when viewed from near the North Pole.

Twelve thousand years ago, the northern coast of the Arctic Sea was habitable, whereas the continental interior was locked in glaciers.

At that latitude, the year approaches the characteristic of a single day: the sun can disappear for long periods in the winter and can stay visible for many months in the summer. Planets can be observed in the long "night" of winter, and the stars can also be observed in their slow precessional drift westward, changing the frame of heaven. Therefore astronomy in the polar regions gives clearer access to celestial phenomena that are otherwise broken up by night and day. For observers closer to the Equator, the daily rising of the sun hides the night sky by day. Evidence for both the precessional changes of the equinox sun and the astronomy of the polar regions can be found in the Rig Veda and later texts, according to the work of B. G. Tilak.[4]

There is climatic support for an unfrozen Arctic Sea twelve thousand years ago, when the Gulf Stream could enter it and warm it. The polar sea could then absorb sunlight, generating moisture that fell as snow to the south, forming glaciers over northern Europe and Asia. Despite the resulting frozen continental conditions, seafood was plentiful and accessible on the Arctic coastline. Social patterns reminiscent of the Neolithic could have been implicit within a language developed there since the Arctic seaboard could have supported a similar social organization.* The original Indo-European language probably evolved over millennia, but matured within a few hundred years; the other root languages were probably maturing about this time as well, but in different regions, in the Age of Leo—ten thousand to twelve thousand years ago.

DIRECTLY VIEWING PRECESSIONAL TIME

To live near the poles is to be immersed in celestial time and hence conducive to forming a profound understanding of the mechanisms of

---

*These reflections are some of those expressed by J. G. Bennett concerning language formation in the Arctic in his *The Dramatic Universe,* volume 4, which prefigured his *Systematics* article *The Hyperborean Origin of the Indo-European Culture* (London: Institute for the Comparative Study of History, Philosophy and Science, December 1963) on the same subject (online at systematics.org and jgbennett.net).

time. The long, dark winter seasons reveal the celestial clockwork, and the year is transformed into a kind of day, as if the sun rises and sets but once per year with an exaggerated twilight around the equinoxes. When the year becomes the primary time period, the rotation of Earth reveals a perfect sidereal clock face in the darkness of winter. One would expect the texts emanating from a polar region to provide a unique view of Great Time, speaking of the interlinked celestial cycles as the days and nights of the gods. The conditions near the North Pole allowed observations, over hundreds of years, of the slow changing of the stars behind equinox and solstice sun in the precession of the equinox. The pole, the center around which the stars rotate, is almost overhead and would be seen to slowly change in a very direct way over hundreds of years.

Once we are aware of how far the framework of stars surrounding Earth shifts in a given number of years, that fraction of the whole zodiac divides into the ecliptic (the sun's path taken as one whole) to form an estimate of the entire length of a full precessional cycle. We can use modern calculation: in 100 years, the equinox sun moves about 1/260th of the whole ecliptic (the sun's path), showing the cycle as about 260 × 100 or 26,000 years. Polar observers would also be able to grasp the circularity of the sun's path after observing the sun travel in a helical arc, from its rise in late spring to its setting in early autumn. Once the sun had set, the planets would be seen to share the sun's path through the sky.

The Greek Hipparchus, credited with discovering precession, was probably not the first. It is likely that Megalithic astronomers first *measured* precession (see chapter 2) and that Hyperborean astronomers were the first to *notice* it, and Neolithic storytellers the first to *explore* its structural implications.*

That a tradition of precessional myth emanates from deep within prehistory was established in 1969 by the book *Hamlet's Mill*.[5] This is made more likely if those living in the far north were the source of this

---

*Hipparchus was the first to follow the modern approach using planetary data and star charts from the Babylonians to *deduce* precession: his method relied on the written notation and arithmetic techniques that were not yet innovated in earlier epochs.

tradition, powerfully expressed in the Finnish epic *Kalevala*. Early on in this tradition, the notion arose that events in the sky might affect the human world in often-tragic ways and that the frame of heaven was in flux, the mental climate and our fate changing with it. If consciousness were varied by precession, like a story with characteristic episodes, then precession would be the story of the world told by the sky and involving a drama taking place there amongst gods, celestial rivers, and a topography with definite poetic potentials. The discovery of precession would have literally rocked the prehistoric world, implying a slow but great change in the whole framework of, by their inference, *meaning*.

SYMBOLS OF PRECESSIONAL MYTH

The symbols of myth appear to have formed a technical language within a storytelling tradition that explained what happened to the sky to make Earth misaligned to the sun, moon, and planets, forming something like a grinding mill. This cosmic mill can be stolen or broken, which can lead to tragic results. The mill was alluded to in *Hamlet,* the tale being told for the last time—very much altered—in William Shakespeare's play.

The Celestial Earth

Hamlet's Mill identifies the scope of such mythic language elements that include the concept of Earth projected onto the sky as the celestial Equator, leading to an above (to the north) and a below (to the south). Throughout the year, the sun travels above and below this celestial Equator. In the Great Year of precession, the constellations on which the sun sits in spring are drowned below the celestial Equator only to be resurrected thirteen thousand years later, in the autumn—a literal *catastrophe* (a word that imparts the concept of stars falling).

The Hindus called the northern ecliptic the Path of the Gods and the southern part of the sun's path the Path of the Ancestors or Demons. These terms conjure an underworld and a divine world, each belonging to a specific age or epoch, and access to these higher and lower worlds is conditioned by the various disruptions in how certain roads or rivers connect to the galaxy in these different ages. The Milky Way forms a distinct river or road that is also caught by precession, since it runs between two fixed points on the ecliptic (the sun's path): between Scorpio and Sagittarius and Gemini and Taurus (see The Galaxy Marks Time, starting on page 33).

The Skambha

Alternatives to the mill include a singular tree (the pole) that falls or is sometimes felled; a churn over which the gods compete, using a cosmic snake to pull against each other and rotate the churn; or a firestick. Waters are released from below, and only the creative fire of the new age can force back these waters to restore a new Earth with different key constellations of stars.

> *And I saw a new heaven and a new earth: for the first*
> *heaven and the first earth were passed away; and*
> *there was no more sea.*
> *. . . for the former things are passed away.*[6]

A single phenomena can be revealed through many alternative views. This is what de Santillana calls a mythic *implex*. The motion of the pole is given a falling nature, a rotating character with alternation. This then leads to the firestick, which gives fire, rising above the celestial equator, to cure the waters gushing up from a hole driven into Earth in the south. There are many such metaphors employed in these archaic worldviews, but it is a singular astronomical reality above in the sky that these metaphors interpret.

Objects or gods can be placed at both the base and the top of the conceptual pole running from south pole to north pole: the turtle form of Vishnu raises up the mill from the waters or it stands on a cube (the New Jerusalem) to signify the framework of space, lying beyond the visible planets, that is changing. In myth, fire must often be stolen from the underworld to establish a new state of affairs on Earth, and this stealing must be accompanied by tasks such as cutting down a polar tree or recovering an object, sometimes fire or "the stone that the builders rejected." Above is the north pole, sometimes represented as another god—perhaps a one-legged god, such as Druva or Vishnu, who holds up the mountain (of the Northern Hemisphere) with a single finger. The same tradition of myth lived on in the New World up until the Spanish Conquest, with similar allegiances to the southern pole and a range of gods, heroes, animals, and metaphors for the altered astronomical viewpoint.[7]

The Frame and Its Colures

The key points in the year are the sun's arrival at four stations, which form a natural narrative structure for the year.

- The spring or vernal equinox, when the rising and setting are exactly east-west all over the globe
- The summer solstice, when the sun's march north comes to a standstill; considered the limit of its ascendancy
- The autumn equinox, when sunrise and sunset are again east-west
- The winter solstice, when the sun's march south comes to a standstill; considered the limit of its decline

In an age, these points are four colures or meridian stripes that pass down the sphere of the stars from the north pole to the south pole, striking the ecliptic (the sun's path) at those points where the sun will sit at these four key times of the year. The four-gated city of the sun therefore gives rise to the four colures of the age, and references to zodiacal signs or indeed any signs within these celestial meridians were often used as precessional symbols for ages that are succeeding their neighbors in one of the sun's four gates. An equinoctal colure, while noted in many ways, is definitively its constellational symbol within a precessional age: when

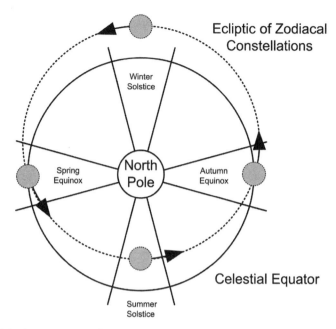

Figure 1.2. A separation of the world parents took place when the sun's path, or ecliptic, and the celestial Equator were separated by Earth's tilt. Thenceforth, these two would meet only twice a year, at the vernal equinox and the autumnal equinox. (This figure is drawn for the Northern Hemisphere.) The drawing of two overlapping circles gives rise to the vesica shape between the worlds of sun and Earth.

we refer to the coming Age of Aquarius, we name the constellation where the sun will soon sit at spring equinox.

The Zodiac of Animals

The zodiac of twelve signs is assumed to be a classical invention of the Greeks, but the evidence in archaic narratives indicate origins further back than ancient Greece. Are the twelve signs of the zodiac arbitrary and merely convenient in number so that we might choose other divisions? We can note that post-Vedic symbolism indicates a system of constellations parallel to those we use today, including similar designations for these and the other constellations away from the sun's path. The Greeks probably inherited the zodiac with precessional lore and did not invent it.

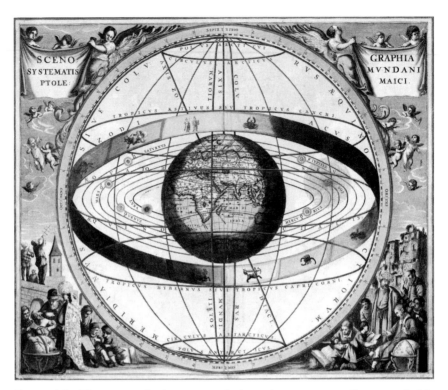

Figure 1.3. The sun and planets travel through some naturally recognizable constellations that mark twelve equal regions through which the sun travels in the year. These constellations are called the zodiac from the Greek meaning "ring of animals." Picture From Andreas Cellarius Harmonia Macrocosmica, 1660–1661.

The choice of twelve constellations is important, because twelve is an organizing power in the world of pure form and number, seen as higher worlds throughout sacred number theory. Twelve is a product of the first two prime numbers, 2 and 3 (being 2 times 2 times 3). Our concept of 360 degrees is perfectly divided by 12 into 30-degree segments and is an idealized version of the 365 days required by the sun to traverse the ecliptic in a year. Like the foundations of a building, the zodiac of twelve houses defines the basis upon which later developments (the system of 360 degrees) was then built, as we see in the next chapter, where the moon divides the year into twelve. A year of 360 days could then have 5 special days.

The Planets or Gods

When combined with planets, the zodiacal signs give a road map for the location of celestial events such as planetary conjunctions, and as these events unfold under a watchful cultural gaze, the entire framework slowly shifts. The shifting of the pole results in the equinox being shunted retrograde (i.e., against the direction of planetary motion). Polar shifting causes changes in the vesica between the celestial Equator and the sun's path (the ecliptic, or zodiac). This vesica was said to have been opened up at the creation of the world, when the world-parents of Equator and Ecliptic—called Shiva and Shakti in the Indian experience—were separated.

The gods are immortal, as are the planets, but they were seen as fallible: giving rise to sins or iniquities that bring about the perpetual falling of the world tree, and causing the whole sky to change its meaning. This perception is accurate, because the sun, moon, and planets determine the precession of Earth. Without the moon, Earth's polar tilt would vary widely due to the planets—and this would be an even greater sin.

Since these two circles of ecliptic and Equator are not aligned, the space between them is like the proverbial crack in the cosmic egg out of which the creative process of the gods arises. Not surprisingly, the vesica shape is paralleled in the female vulva as a source of new life and sexual creation. The stage was thus set for a vast literary parallel between human creativity, sex, sin, divine creativity and the creation of a perpetually changing world of time.

The Galaxy or Cosmic Road

The division of the ecliptic by the great circle of the galaxy, or Milky Way of stars, is the body of stars to which the Sun belongs. It therefore represents directly any pathway to a supreme creation and its purposes (if any). If the unity of sky and universe was subject to a creative act, then what we see on Earth is this Milky Way running through it—a path that slowly rotates with precession to alter the spiritual topography

Figure 1.4. The Milky Way, or rainbow bridge, crosses the ecliptic as it rises high to the north, adding a third galactic framework to that of the zodiac and Equator. We can see the center of the galaxy in the bright stars and a dark rift on the right, toward the constellation Scorpio. Picture by Dan Duriscoe for the U.S. National Park Service.

of each age, in its galactic connection to the framework of the four colures.

The galaxy is most aligned when two of the four points, either the solstice or the equinox colures of an age, coincide with the galaxy to form the connection to the higher worlds that the galaxy represents. The Age of Leo enjoyed the center of the galaxy behind the summer solstice sun and, therefore, its anticenter behind the sun at winter solstice. Today, the opposite is the case, and perhaps this lies behind the expectation of change associated with the Age of Aquarius, headlining as the end of the Mayan long count and fifth sun, or age, in 2012.

The galaxy is therefore a significant remnant of creation and deviates from the sun's path: Helios is said to have lent his chariot to Phaeton (whose name means "shining"), who drove it for a day in reckless fashion to near destruction of the celestial Earth.

The unity of the celestial sphere and its eccentricities were therefore conducive to painting highly unified thoughts about the cosmos and the human role within the creation by using a technical language of symbolism that could be read into the phenomena themselves. This is thinking about a phenomena using language and behind it lies a profound possibility: that the mental climate for our thoughts is somehow defined by the stars and their orientation seen from Earth. Since stars are in fact the primary players in the creation, then this insight would be highly integrative and truly religious in proposing a framework that reconnects the human mind to its origins within the creation.

The structure of the story form, as a circular narrative, is directly available in the nature of the ecliptic circle punctuated in four places by the points at which the galaxy integrates with four different ages—those of Leo, Taurus, Aquarius, and Scorpio—which then lock with the rotating fourfold colures of equinox and solstice events within the year. The colures form part of an apparatus that selects areas of the zodiacal stars: those stars that lie in the plane of the solar system that itself selects just one band of the whole starry sky.

If we use as an archetype the sun's path in the year, then the Age of Leo is high summer of precession, the Age of Taurus is autumn, the Age of Aquarius is winter, and the Age of Scorpio is spring. The latch of the story is then naturally formed as both an origin and destination in Leo; and the turn, as the middle of the story, occurs in the Age of Aquarius, which is imminent today. The present moment of the precessional story is the location of the galactic center as it slowly moves relative to the fixed colures of the year, like an hour hand on a clock of the Great Year of twenty-six thousand years.

The cycles of the planets are echoes of this same nature; each planet is given a different character as a god who travels within the circle of the animals, which involves climbing a hill of heaven or descending into the underworld below the Equator. The language is effective; it does not descend into functional descriptions, as does today's astronomy, with its frame of reference abstracted into solar system or galactic coordinates. Indeed, the storytellers came to a natural language in which the symbols themselves told the story. Consciousness at any given moment was an amalgam of the location of the cosmic centers in their progress along the great rivers of heaven that formed a celestial Earth.

Once the cave of heaven and its mill were sketched out as to their major characteristics, the path was open for transmission through myth within the cultures that came after this. More could be known, but additional knowledge would require the more detailed observations achieved, most plausibly, by the Age of Taurus—after one of the four galactic meeting points and in which the megalithic peoples would build monuments that give a still more detailed view of the same story of the mill.

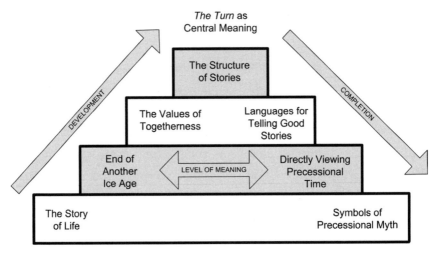

Figure I.5. The concepts laid out in this chapter are presented in a step-wise, "pedestal" format, which is a traditional alternative to the circular or ring-type compositional structure. This diagram demonstrates levels of meaning in the opposite halves of the story's structure. It also shows how such a concern with compositional structure can contribute to the understanding and clarity of the story as a whole. Literary specialists think that such structures enabled an oral tradition of storytelling to maintain the story reliably, since the loss of form, and therefore content, could be checked by those trained in this art. Even today, much can be learned from the Gospels, for example, as to the effect of editors upon what once was a series of tightly crafted literary structures. Religions tend to ignore the fact that, with respect to the scriptures, specialists composed what they purported to be a spontaneous reporting of events. Yet the scribes (see chapter 3) were also those trained in the canon of spiritual symbolism that descended from the precessional myths, which were solidly established in the lives of peoples by the Age of Gemini.

THE GALAXY MARKS TIME

THE BASIC THESIS OF cosmology is that the small is somehow shaped by the large. These can be listed as:

- The moon, which follows its own unique course (the moon god)
- The Earth, which is tilted and spins once a day beneath our feet (an earth mother)
- The sun, which is the center of the solar system (the sun god)
- The planets, which surround Earth in a highly sophisticated set of interrelated motions (the gods)
- The galaxy, which surrounds Earth with a ring of millions of stars, but whose center lies in the boundary of the constellations of Sagittarius and Scorpio (the cosmic serpent)

Precession occurs due to Earth's tilt, causing the equinox sun to move slowly backward through the zodiacal signs. In this process, the galactic center is the only cosmic concentration that moves continuously throughout the cycle of precession. If this highest level of scale, representing millions of stars, is the highest visible manifestation of the whole of creation, then signifying the galaxy as a representation of the spiritual world is quite natural and may be why precession underpins traditional wisdom.

It is unlikely that the galactic center should lie near the ecliptic, because the galactic center could lie at any point on the great circle of the galaxy. The fact that it lies near the sun's path once during the year requires the plane of Earth's orbit to coincide with the center of the galaxy (within 5 degrees).

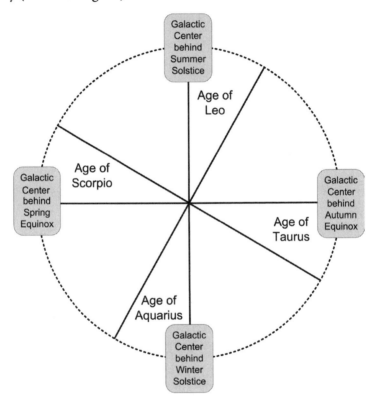

There are four ages of the twelve precession ages that start with an alignment to the galactic center. Between these are precessional seasons. The present age, around 2012 CE, is inaugurated as the Age of Aquarius. Storytelling appears to have started at the point opposite today, crediting the Age of Leo thirteen thousand years ago with language development. The Megalithic Age of Taurus was inaugurated sixty-five hundred years ago, and the last Age of Scorpio could well have initiated the cave art of the late Paleolithic period.

The consequent orbiting of the galactic center within precession can be shown as a circle in which there are four cardinal points of great interest (see figure on page 34). Each of these occurs when one of the solstice or equinox points of the sun in the year play host to the galactic center, sitting behind the sun. Seen in this way, we can note a peculiar similarity to the solar year: the order of the precessional round follows the natural form of the solar year. The galactic conjunctions repeat the sequencing of the four gates of the sun *on a higher level,* exactly as if the galactic center is a higher form of sun for Earth. Any observance of the galactic cycle could easily be mistaken for a form of solar worship, and therefore a precessional cult can be confused with a solar cult. In aligning the entrance passages of burial mounds to the solstice points, builders of burial mounds probably tried to connect to the galaxy, the traditional celestial path to the higher worlds for their cults of the dead.

When the galactic center sits on one of the four points, then the galaxy also touches the opposite point, solstice to solstice or equinox to equinox.

The symbol of three lines crossing represents perfectly the times when the galaxy crosses the equinox points of the year. Hertha von Dechend believed that the labrys, or double-headed axe, was associated with the precession of the equinoxes. From Catal Huyuk Shrine VI.A.66 after Mellaart.

At the equinox points, the three great circles of the Equator, ecliptic, and galaxy can coincide, because Equator and ecliptic already coincide when daytime *equals nightime,* as the word *equinox* itself means. In the cycle of an individual day, these are the points of balance of light and dark, or twilights, sacred to Hinduism. The transitions from one age to another are also called twilights, the word meaning both double or half-light.

The galactic center, therefore, hits each one of the four primary points in the solar calendar just four times in twenty-six thousand years, in the correct seasonal order. The period of sixty-five thousand years, commonly referred to as a precessional season, has added meaning when these galactic conjunctions are taken as four singular moments in the cycle.

Today, we are at the bottom of the four points, because the galactic center stands near the winter solstice. Time can be said to move with the galactic center, whereas the galaxy moves in our experience of time. This environmental fact, caused by the inevitability of the precession, could hide the fact that our present moment is always structured by cosmic factors.

The galactic center could well provide the spiritual framework of an age and make up the conditions for the development of new modes of consciousness. Our minds have been conditioned by our culture to dismiss the possibility that the cosmic environment provides a learning environment for consciousness. It is therefore difficult to realize the possibility that the galactic center is transforming life on Earth, even though it is the cosmic concentration of the greatest scale that envelops Earth and the solar system.

Shri Yukteshwar expressed it this way: "After [13,000] years, when the sun goes to the place in its orbit which is farthest from *Brahmah,* the grand centre, *dharma,* the mental virtue, comes to such a reduced state that man cannot grasp anything beyond the gross material creation. Again, in the same manner, when the sun in its course of revolution begins to advance toward the place nearest the grand centre, *dharma,* the mental virtue, begins to develop; this growth is gradually completed in another [13,000] years."[1]

We will see the significance of such cosmic concentrations in changing laws in chapters 5 and 6, but such effects are not through causal influence. Instead, the framework of the universe itself, altered by a cosmic concentration, affects the subjective and qualitative aspects of life: embedded within a mill and its frameworks that subtly alter "that which can happen." In this sense, therefore, the gods are actually present, though they appear far away in the sky, and cosmic concentrations need to be distant to set up conditions on Earth without destroying Earth. This felt presence of the gods has diminished today, a fact much bemoaned in myth as the withdrawal of the gods. The seeds of this withdrawal were initiated in the Age of Taurus, during which the Megalithic civilization arose, when a new activity began: investigating the time world of the celestial mill.

Chapter 2

MEASURING THE
COSMOS IN TAURUS

THE EFFECT OF THE seasons, sun, moon, and the periodicities of the planets as experienced on Earth give celestial time a definite pattern. Stone Age stories about the sky could be extended only by moving on, by gleaning a better understanding of the exact order of celestial events from the stable base of Neolithic agriculture. Such a realization led to the cultural advance that we now call the Megalithic period, during the Age of Taurus. The Megalithic period evolved a unique approach to measuring and calculating, a method that was quite distinct from that of our present science and mathematics, and these unfamiliar methods were recorded in stone. Their monumental architecture, required to measure the numerical properties of celestial time periods, developed from some very simple ideas about counting and geometry.

THE DEVELOPMENT OF THE DAY-INCH

The primary unit of measure used in the Megalithic period was the day itself. Counting days allowed significant achievements: for example, the lunar month can be counted over two entire periods (59 days). At 29.53 days, the lunar month is nearly 29.5 days, which means that two months equal 59 days, to nearly 1 part in 1,000. This provides

Figure 2.1. One of many alignments at Carnac, Brittany, where, for unknown reasons, many rows of stones exist at about 23 degrees east-northeast through the landscape

therefore a very accurate approximation without much sophistication in measurement, a recurring theme in megalithic astronomy, where achievements were made possible through fortunate astronomical circumstances.

Such counts are found in the Stone Age as sets of notches or other marks scored on bone. These marks later came, by the megalithic, to be made with equal spacing using a size similar to our present inch—a digit related to the width of a thumb. An inch measure has many uses; it is conveniently small, and a great number of them, used in a longer count, generate lengths that can be used to construct geometrical structures through which time periods can be compared as objective ratios. The adoption of a uniform length for each day also allowed the detection of fractions of a day when these counts were employed geometrically: 1/8 inch could be visible as 1/8 day (three hours). The resulting system

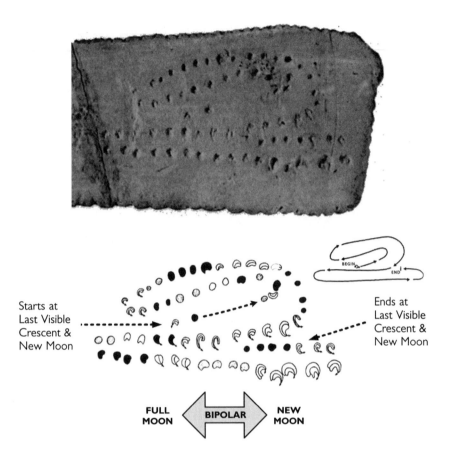

Starts at
Last Visible
Crescent &
New Moon

Ends at
Last Visible
Crescent &
New Moon

FULL
MOON

BIPOLAR

NEW
MOON

Figure 2.2. Engraved and shaped bone plaque from the rock shelter site of Blanchard (Dordogne) from the Aurignacian period, about thirty-five thousand years ago. This was one of many bones that Alexander Marshack, in his book *The Roots of Civilization*, interpreted as being based upon the phases of lunar months.

of *metrology* allowed the study of the cosmic time ratios, such as those between the solar year and lunar year (of twelve lunar months). This activity appears to have been the precursor for our system of metrology—a system based on ratio—inherited as the historical measuring systems used throughout the world.

The four years between identical sunrises on a horizon mark could be counted as 1,461 day-inches, and the division of this by 4 (by folding a marked rope twice) results in a 365.25 day-inch year. Such direct measurements contrast with our present methods, in which measurements are abstracted from a measuring apparatus using numbers—that is, through the use of numerical notation. Instead, Megalithic astronomers maintained a direct connection between the apparatus and the measurement by using geometrical techniques based upon inch counts.

This approach enabled powerful calculations using exactly the type of geometries we find at Megalithic sites. These frequently employ right-angled triangles, which allow the comparison of day counts without using our trigonometric methods: the difference between the two longer sides of a right-angled triangle presents the cosine function of our trigonometry. Metrology, a technology involving measured lengths, could incorporate time measurement into such triangles once a constant day-inch is used when counting.

STUDYING THE LUNAR AND SOLAR YEARS

The most obvious time proportion to study is the difference between the 12 lunar months of a lunar year (354.367 days) and the days in a solar year (365.2422 days), which is slightly longer. As already stated, day-inch counts can give reasonable results for the length of the solar year and lunar month, at least as good as 365.25 day-inches and 29.5 day-inches, respectively. The lunar year is then 12 × 29.5 days, or 354 day-inches long. However, a longer count of 3 lunar years (36 lunar months) yields an extra day to the count—1,063 days—and this is then ten times more accurate than a 29.5-day lunar month, being accurate to almost 1 part in 10,000. It is only 2.5 hours short of the correct time for 36 lunar months (1,063.1 days).

Brittany is one of the oldest Megalithic regions, and the region around Carnac (dated around 5500 to 5000 BCE) has many complex monuments that relate to the sun and moon. At the latitude of Carnac, the sun's rising and setting points at winter and summer solstices has the strange geometric property of forming the first Pythagorean triangle relative to the east-west axis. Many such triangles, with side lengths of 3, 4, and 5 units long, were constructed at Carnac, and within them, two sides define a right angle aligned to the cardinal points. The longest side (5 units long) was then aligned to the direction of the sun at sunrise or sunset during both solstices.

The presence of megalith builders at this latitude demonstrates that the simplest of all the whole number triangles was understood at that time and that it lay at the root of these builders' geometrical thinking. This triangle affirms their choice of this latitude based on this simple triangle, as it has the practical application of easily establishing sightlines to the solstice sunrise and sunset.

Such a triangle can be constructed using a rod of any length to mark three sides of a triangle (3, 4, and 5 units in length). In fact, such whole-number triangles are best constructed on the ground using ropes,

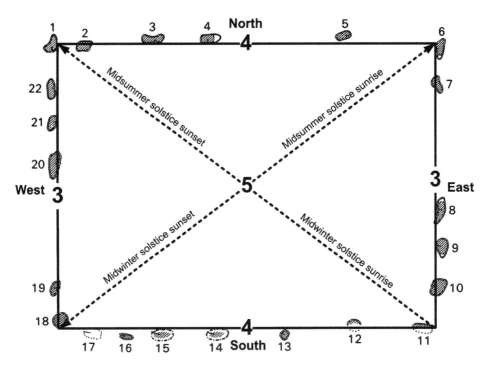

Figure 2.3. The usefulness of 3-4-5 triangles at Carnac in aligning to solstice sunrises and sunsets. One of the clearest examples of this geometry can be found within the Crucuno rectangle. Alexander Thom's 1973 survey of this rectangle, despite having many of its stones disturbed, fallen, or ambiguously re-erected, demonstrated that 3-4-5 triangles were regularly incorporated in the Megalithic structures at Carnac.

a method in which the right angle makes itself during the process of manipulating a regularly knotted loop of rope. Since day-inch counting lengths can be compared using a right-angled triangle, a pegged rope ensures that an accurate right angle can be formed in triangles whose sides were not 3-4-5, as shown in figure 2.4. This technique could then be used to provide the right angle for triangles with longer sides based on day-inch counts.

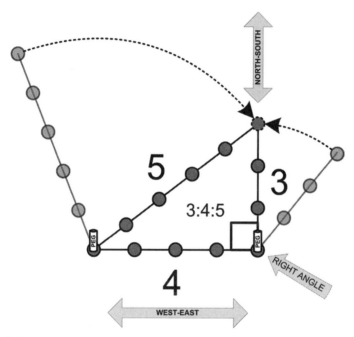

Figure 2.4. A rope with twelve equal divisions comes together to form a 3-4-5 triangle and an accurate right angle. Any Pythagorean triangle made of whole numbers on each of its sides was used to build right-angled triangles with day-inch counts on their two longer sides, allowing Megalithic astronomers to geometrically study celestial time ratios.

Having established that the solstice sun always shone on the longest side of a 3-4-5 triangle at the latitude of Carnac, the Megalithic astronomers chose the solstice sun alignment to begin a count in day-inches. In 2009, Robin Heath discovered an inch-encoded sightline at the site known as Le Manio, within a construction called the Quadrilateral (see figure 2.5). This discovery supported my theory that day-inch counting lay at the heart of early Megalithic astronomy.[1] The length of the line in day-inches is found to equal 3 solar years (1,095.75 day-inches), or as we would say today, this line is three years long (we still refer to time as having a length).

Figure 2.5. The day-inch counting line defined by the midsummer solstice at Le Manio Quadrilateral. Photo of the site from the northeast. Photo by Robin and Richard Heath.

In a 3-year count, it is natural to mark each day-inch within which a full moon occurs. Thirty-six such full moons result over 3 lunar years, if the count begins on a full-moon day, and the extra length required to reach the end of 3 solar years reveals the difference between 3 lunar and 3 solar years: just over one further lunar month.

Forming a Soli-lunar Triangle

The Quadrilateral at Le Manio was built using small megaliths that touch, so as to form low walls called *kerbs* punctuated in places by gaps, and some gaps were evidently symbolic as with the (solsticial) "sun gate" from which the three-solar-year, day-inch count began. A triangle was formed from the counts by dropping one end of the 3-lunar-year count

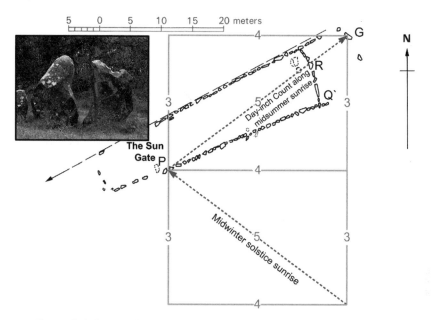

Figure 2.6. Photo of the distinctive sun gate at Le Manio, including stone P, where the counting began

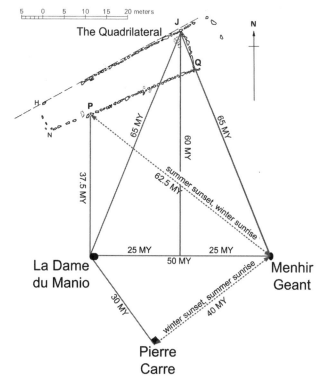

Figure 2.7. The local geometrical interpretation of Le Manio using 3-4-5 and 12-13-5 triangles within an integrated scheme. The Association pour le Connaissance et l'Etude des Megalithes, or ACEM, is continuing the work that was done over the last 20 years at Carnac.

until it formed a right-angled triangle that traveled along the south-ernmost kerb of stones that now make up the Quadrilateral (see figure 2.6 and the box *How the Megalithic Yard Was Defined at Carnac's Le Manio*—below—for a more detailed description). This enabled the relative proportion of the two types of years—solar and lunar over three years—to be viewed as a geometric ratio (which is what the slope angle of any right-angled triangle represents). In math-speak, a solar hypotenuse above a lunar base results in the ratio between the two lengths, which today we call the cosine of the angle between the two sides, and the base ÷ hypotenuse = cosine. This angle is an invariant property of the sun and moon, and so, wherever this angle or triangle arises, this astronomical reality is presented.

How the Megalithic Yard Was Defined at Carnac's Le Manio

The Le Manio site probably represents early and later developments, and we can now propose a timeline for the earliest part of this development.

1. A suitable line was established that ran along a sightline aligned to the midsummer solstice. Stone P was erected to indicate the starting point for a day-inch count, and the groove in stone G is aligned to midsummer sunrise seen from sun gate P.

2. A 3-solar-year day-inch count was started that progressed from sun gate P toward stone G. This could have used a measuring rod (ruler) designed to accommodate a counting marker.

3. Full moons could have been shown in the count (as with an inch marker made of quartz) as the count progressed to its 3-year end. Stone R was eventually installed after three years and dressed to provide an accurate reference to the end of the count.

4. The distance between the 36th full moon (lunar month) and the end of the 3-year count equals 32 5/8 inches. This length be-

came the standard megalithic yard of 2.718 feet used later within the Le Manio site and beyond. A standard rule of this length could then be constructed, possibly calibrated as a ruler using a set of inch markers, grooves, or notches.

5. A triangle was formed to present the differential length between the solar and lunar counts. A second rope, whose length corresponds to 36 full moons (lunar months), was attached to the left end of the original rope (at sun gate's point P; see figure 2.7). The new rope (36–full moons long) is then pivoted downward from the joining point at P until its end stands below the end of the longer (the 3-lunar-year-count) rope. The two rope ends define a third side that forms a right angle between itself and the count for 36 months. A 12-section rope would have helped achieve such a right angle using a 3-4-5 triangle, as discussed above.

6. Stones R and Q (see figure 2.8) were only later prepared and installed to mark the end of the 3-solar-year count and the end of 3 lunar years, respectively, and to set the triangle in stone down two of its sides as part of the Quadrilateral. The original count forms the hypotenuse between the sun gate's point P and stone R, between which the count (the hypotenuse) can still be measured.

Whenever a megalithic yard was not available, a 3-year day-inch count could provide it from a standardized inch using the above geometrical procedure.

The slope angle of a triangle's longer side relative to its base causes any sub length of the base to correspond to a longer sub length above it on the longer side (the hypotenuse) according to a fixed proportion. A lunar-month count on the base therefore points to the longer mean solar month (1/12 of a solar year) on the longer side. This difference in

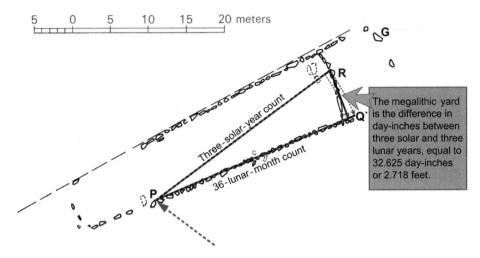

Figure 2.8. The triangle constructed at Le Manio from counting 3 solar years and dropping the (initially) parallel count of 3 lunar years in order to create a right-angled triangle.

the growth of a hypotenuse relative to its base results from the proportional growth in length between the two sides. This geometric structure enabled a Stone Age culture to make scientific discoveries through constructing metrological and geometrical tools on the ground.

Once they generated such a triangle, the builders could see the proportionality between lunar years and solar years: lunar months on the base translate to solar months on the solar count hypotenuse. This naturally projects the division by 12 of the lunar year (by lunar months) into an equivalent division of the solar year (by 12 solar months). These mean solar months are not marked by any directly visible phenomenon, as they are with the moon (where a full moon divides the lunar year into 12), but this triangle naturally presents the division of the ecliptic into a zodiac of twelve equally spaced regions through which the sun travels through the year. It is likely that the twelvefold division of the sun's yearly journey through the stars was triggered by this clear projection of the lunar year onto the solar year and that this was discovered when the two time periods were compared using a triangle in the Megalithic.

This expansion of the lunar month into a solar month generates the difference between the shorter, lunar base and the longer, solar hypotenuse after 36 (3 × 12) expansions. At Le Manio, the count appears to have generated a difference between 3 solar years and 3 lunar years of 32 and 5/8 inches, or 2.72 feet, which was the length discovered independently at Carnac by Alexander Thom, and was the length he named the *megalithic yard.*[2] It is important to note that this yard is shorter than an English foot yard, which is 36 of the same inches.

The builders therefore appear to have used a day count to establish their megalithic yard, defined as the excess of 3 solar years over 3 lunar years in day-inches. This unit of length was then employed for subsequent constructions, such as the 3-4-5 triangles, using measuring rods or ropes based upon such day-inch counting.

THE METER AND THE MOON

My brother and I returned to Le Manio at the March 2010 equinox in order to see how my theory of day-inch counting would correspond to an accurate survey of the Quadrilateral.* We had realized that the solar-lunar triangles (there is also one for 4 solar years) were calibrated in day-inches rather than the expected megalithic yards because they were real day counts. A number of additional facts emerged that further explained the unusual design of this megalithic monument.

The third side of a soli-lunar triangle for 3 years has a length exactly equivalent to 9 lunar months of day counting. The 36 months of the base can therefore be divided by this length exactly 4 times, because 36 ÷ 4 = 9. This provides a geometric method to reconstruct this triangle over any period, not just 3 years, to generate day-inch counts without actually counting the days.

By generating 4 squares of equal side length equal to 1/4 the base, the size of the diagonal (the triangle's third side or hypotenuse) is then the solar-year count, which is 12.369 lunar months long (as a

---

*The survey details and its full set of conclusions are published at precessionaltime.com and skyandlandscape.com.

day count)—the square root of (a) the square of the base of 12 months plus (b) the square of the third side of 3 months. (Today, we use the Pythagorean theorem, adding the squares of 12 and 3, 144 + 9 = 153, which has a square root of 12.369 months.) The achieved accuracy of this geometrical diagonal relative to a day count for the given number of solar years is 1 part in 20,000, which is very accurate. The geometry of four squares therefore exceeds the practical accuracy of any actual day count; it is within 45 minutes per solar year, and can thereafter replace day counting as a method for reproducing this triangle.

The four squares geometry must have become a portable procedure for the reproduction of the soli-lunar triangle (both its slope angle and actual day-inch count) for any chosen number of solar years. Its reproduction probably required the evolution of a standard measuring unit with which to build this geometry—a unit more convenient than 29.53 inches. Such a natural unit emerged during our survey when we noticed that 36 months measured exactly 27 meters on the base for 3 lunar years. The modern meter is 36/27 (4/3) of the day-inch count for a month, because 3/4 meter equals 29.528 day-inches. Three meters—118.11 inches—then becomes a useful measure equivalent to a 4-lunar-month day count. This

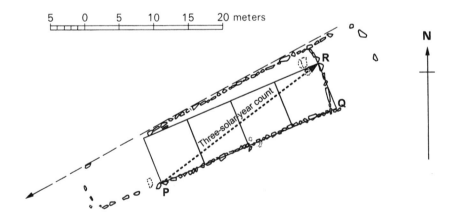

Figure 2.9. Four-squares geometry, with a diagonal that represents the solar year. Once discovered, this geometry leads to greater accuracy than day counting and provides a portable geometric device for reconstituting the soli-lunar triangle's day-inch count.

3-4 relationship between the month and the meter plays out between the 3-year and 4-year triangles at Le Manio, because the 4-lunar-year base is then 36 meters, the number of months in 3 lunar years.

The 4-year triangle at Le Manio shows the meter's utility very clearly, because it is made up of 4 12-month squares, each of which is 3 meters × 3 meters in size. Whereas the eastern corners and southern kerb, defined by the 3- and 4-year triangles, can be read in meters as well as inches, lunar years are more easily counted as meters that equal 4/3 months, because a lunar year is 12 months long and 9 metres long in day-inch counting.

Thus the Quadrilateral is a perfect symbol of the essential sun and moon ratios, built using the number symbolism of metrological lengths. The Quadrilateral appears to have been built exactly at the latitude at which it was possible to fit these cosmic ratios, as triangles, between the midsummer sunrise and the east. This use of latitude implies that the builders had already mastered the additional skills of determining latitude and surveying. It also implies that these day counts were not originally conducted at Le Manio, but had led to a search for the latitude at which these two triangles would fit in this way. The monument at Le Manio was probably built as a recapitulation of early Megalithic knowledge, written in stone as a perfect statement of what the builders knew and how they had come to know it at some time prior to 4000 BCE.

The use of counting lengths and geometry was the ideal mechanism for the people of the Late Stone Age to have developed numerical thinking, prior to, and in preparation for, the development of number notation and its associated mathematics. Such use enabled a Megalithic science that could tell the story of cosmic time—prior to the known history of mathematics.

THE EFFECT OF AN INTELLIGIBLE COSMOS

It is clear that the Megalithic astronomers pursued very simple strategies that successfully unlocked some remarkable characteristics in the relative time periods of the sun and moon as seen from Earth. Their methods required the use of geometry and the metrological skills that are found at megalithic sites as part of an ancient wisdom thought to have existed in prehistory and probably mythologized as the story of a lost high civilization called Atlantis.

Ancient scientists had probably not detached themselves from the religious concept that the universe was a creation. They could explore phenomena as meaningful expressions of order, and in their view, this would bring them into contact with an intelligence that had created this order. This belief, within precessional time, proved to be a developmental process for their consciousness and its evolution. It left behind our historical metrology still based upon the inch and foot, though ancient scientists' astronomical discoveries soon became garbled or lost altogether to our histories.

The universe, it seems, has something to say, and this can make us more intelligent, as in the principle of intelligibility within the sun-moon system of time. Consciousness is only truly available in the confluence of the intelligible and the intelligent. This requires that a world should be intelligible and life arise within it that can be intelligent, as in the form of the story in which the blending of these two create a narrative structure for consciousness.

If a stick is 19 units long, and we point it at the North Pole from the ground near Stonehenge, in England, the base of the triangle created is 12 units, and the excess of the hypotenuse over the base is 7 units. This means that the megalithic yard, royal cubit, and English foot relationships of 19/7, 12/7, and 7/7 feet, respectively, were all implicit at the latitude where these units are found within monuments in southern Britain.

Such synchronicity between culture and cosmos suggests a helping hand from Nature in that day counting can achieve metrological ratios and simple geometries that can effectively describe the sun and moon. This was also a work that only became possible *within a Neolithic revolution*. If the cosmic ratios are intelligible as geometrical ratios and interrelated periodicities, then an early human culture will always be able to

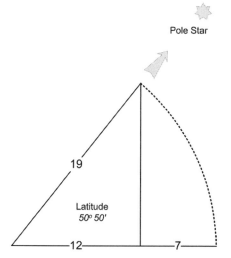

Figure 2.10. The strange coincidence that the key ratios of the megalithic yard (19/7 feet), English foot (7/7 feet), and royal cubit (12/7 feet) are implicit in the polar sightline at 50 degrees, 50 minutes of latitude, just south of Stonehenge. The perfect megalithic yard would correspond exactly to the ratio between three lunar and three solar years. We call this the Astronomical Megalithic Yard, or AMY, and its value is almost exactly 19/7 feet long.

achieve an exceptional degree of knowledge by using the direct methods of the Megalithic.

Two theories could account for how such triangles work so closely with the structure of astronomical time, both suggesting that the cosmos has a designed nature. The first theory proposes that these geometries were the nominal targets for a design process natural to the creation of planetary systems in which life like ours can arise. The second theory proposes that the organization of time then became intelligible, through its simplicity, thus ensuring that early astronomers could develop geometries and ratios, allowing a glimpse of the type of intelligence responsible for the creation.

The convention is that angels or a demiurge are beings rather like us and that they design as we do. It is more likely, however, that there is a form of intelligence within the cosmic structures that arrives at certain types of solutions within the number field. As Johannes Kepler suggested, God geometricizes. Such an intelligence would be expressed within cosmic phenomena. Though the present paradigm is of intelligence coming as a result of biological evolution alone, our level of development is probably only a prerequisite for receiving and acting out a common cosmic intelligence.

Though the theory of intelligent design is disapproved of in modern science—because material causes come about through blind forces and natural laws—this does not mean that a theory involving intelligent design is inherently disreputable or that no new structural laws remain to be discovered. Consciousness and higher intelligence are not purely objective; there is always a subjective component that relates to moments of intelligence in our lives that we cannot repeat, yet a scientific experiment must be repeated in order to prove a result. In considering whether precession affects human history, the Megalithic peoples apparently discovered and were affected by the near-perfect geometrical coincidences they found, ready made, between key time periods. They demonstrated intelligence in comprehending the order of time and being able to realize these relationships in monuments that measured and made them manifest.

Limiting our science to repeatable causation only limits any theory of how time's coincidences play a role in structuring actual situations, and such a science has made the Megalithic achievements unintelligible. The intelligibility of the natural world requires that higher intelligences act through special opportunities that, though seeming impossible or unlikely beforehand, are routinely accepted once they have happened.

THE LONGER CYCLES OF THE MOON

Sometime after the work at Le Manio, Megalithic astronomers discovered that after 19 years, the sun, moon, and stars return to almost an identical position relative to each other. This fact, which we call the Metonic period (after a Greek named Meton) was less easy to observe than it is to predict based upon the soli-lunar triangles these astronomers had discovered. It is possible, for example, to build a triangle that shows exactly 7 months separating the 228 months of

19 lunar years, and the 235 months equaling 19 solar years.*

On the other hand, similar eclipses recur every 223 months, a cycle called the Saros, which lasts 18 years plus 10.875 days (the excess in day-inches for one year). There are exactly 12 lunations in difference between the Saros and the Metonic cycles, which means it is true that the eclipse cycle is quite synchronous with the sun and moon cycle of 19 years. Some basic facts about why eclipses occur reveal another type of year that punctuates lunar and solar years: an eclipse year.

The moon in its orbit (angled to that of Earth) must cross, in two places, the sun's path. When the sun is at one of these nodes, a crossing by the moon causes a solar eclipse at the sun's position or a lunar eclipse at a position opposite the sun. The sun can visit these eclipse nodes only twice in its yearly round; therefore there are two eclipse seasons in a year, but because the moon's nodes move retrograde along the sun's path, the sun meets a given node once every eclipse year of 346.62 days (18.618 days short of the solar year).

This third, eclipse, year occurs 19 times within the Saros cycle, which is itself 12 months short of the Metonic cycle of 19 years, which is 5 months more than 19 lunar years. The cycles are therefore interrelated and have in common the number 19. The regular punctuation of suitable conditions for lunar eclipses in particular means that it is very easy to deduce that the number of eclipse years in the Saros cycle must exactly equal 19 eclipse years.

An eclipse year is therefore 223/19 (or 11 14/19) months in length, which equals 346.6 days, an accurate figure that can be refined through long counts between many eclipses. Based on this estimate, eclipses can be predicted by actively tracking the location of the nodes on the zodiac. This allows the prediction of eclipse conditions without knowing that an eclipse occurred one Saros period ago. A suitable device for

---

*If 7 months separate the two periods, then the excess is 228 inches—if a mean solar month is 1 inch larger than a lunar month. If we use feet as a measurement, this translates into 19 feet, and the megalithic yard for the month is thus 19/7 feet, the astronomical megalithic yard (AMY) found in British Megalithic construction.

this prediction has been found in the form of the Aubrey Circle of holes around Stonehenge, through which, using markers for the sun, moon, and lunar nodes, a sidereal simulator of the sky can be constructed.[3]

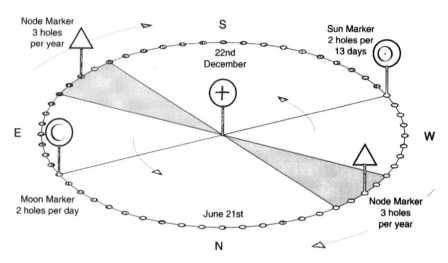

Figure 2.11. The application of Stonehenge's Aubrey Circle of 56 postholes as a circular simulator that uses simple rules to track the sun, moon, and lunar nodes as these progress through the ecliptic zodiac. The slight cumulative errors this simulator generated could be corrected when an actual eclipse is slightly late or early. The sun's path during the year and the stars behind the moon also make this "clock" a sidereal map of the zodiac "in the round."

The triangle for the lunar nodal period, viewed as a 346.6-day to 365.25-day triangle, displays the interesting feature of an intermediate hypotenuse (at the point that divides the third side according to the golden mean; see figure 2.12) close to the lunar year's length.

This 18.618:19.618 triangle, found in the Megalithic structures in Brittany and Britain, is the Type B–flattened circles identified by Alexander Thom. See figures 2.13 and 2.14 on pages 58 and 59.[4]

THE MOON'S RELATIONSHIP TO PRECESSION

The precessional cycle of equinoxes does not appear to be related to the moon's nodal period of 18.618 years. Yet the nodal period would

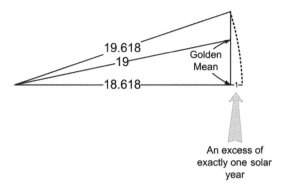

Figure 2.12. The triangle for the lunar nodal period where N = 18.618: as with other N : N + 1 triangles, the difference between the base of N units and the hypotenuse of N + 1 units is very significant—as with the megalithic yard being the difference within the three-year triangle. In this case, the difference emerged from the fact that there are 19.618 eclipse years in 18.618 years, the nodal orbital period. But it is also true that these two types of year, the eclipse and solar years, differ by 18.618 days, because the nodes move one day of solar motion in the time that the sun moves 18.618 days. The 19-year Metonic cycle then sits at the golden mean point on the third side (a property of this type of triangle with non-integer, fractional parts equal to 0.618).

divide into the length of a precessional age if it were 1/15 longer, corresponding to a pure half tone increase of 16/15. In order for this to be true, the length of an age would be 2,145 years long and precession 25,740 years long. Because the modern scientific estimate for precession is 25,750–25,770 years, 1/12 of this is within a few years of the modern estimate; thus the moon's nodes appear numerically related to the precessional cycle, in addition to the moon stabilizing Earth's tilt and hence precession itself.

The figure of 2,160 years for an age, adopted since Plato's time, is based upon an ideal harmonic form (see chapter 3) in which the resulting precessional cycle of 25,920 years was a number amenable to harmonic analysis (involving only the three lowest primes—2, 3, and 5). Such an ideal number is called *canonical*.

It has long been held that the period between Jupiter and Saturn conjunctions, called the Trigon period, was the natural subdivider of precession, even though the evidence appears weak when aiming for the figure

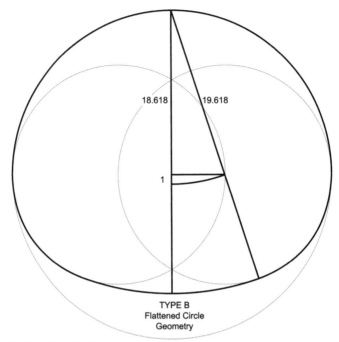

18.618   19.618

1

TYPE B
Flattened Circle
Geometry

Figure 2.13. The Type B–flattened circle is a geometry generated by dividing the circle's diameter into three parts. The lower semicircle is then replaced by arcs radiating from these centers and, between these, by a diameter arc that joins them. The triangle formed by the arc between the two inner centers is the right-angled triangle for the nodal period, as in figure 2.14, a fact realized by Robin Heath as proof that Megalithic astronomers investigated the nodal period geometrically, once they had elucidated this ratio as a triangular form.

of 2,160 years for an age and 25,920 years for the full precessional cycle. Platonic assumptions aside, the Trigon period's duration, 19.859 years, is 16/15 of the lunar nodal period of 18.618 years. The Trigon period is therefore related to both the modern estimate for precession and the moon's nodal period through its relation of 15/16 to the Trigon period.

The Trigon period can therefore be viewed as the natural type of minute for the clock face of the Great Year—and its 12 signs are like the hour divisions to which the hand of the precessional *present moment* points; making the modern clock face is a perfect metaphor for precession. There are 108 Trigon periods in an equinox age, which is a canonical number (9 × 12) traditionally associated with the moon.

The moon, key to stabilizing Earth's precession, is therefore found

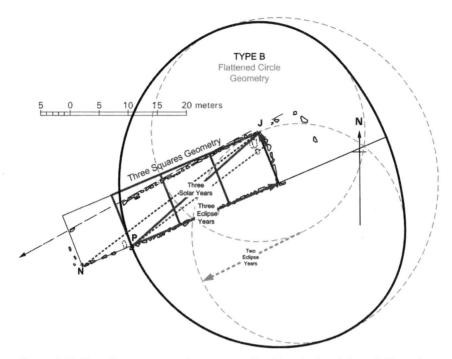

Figure 2.14. Type B geometry in the context of Le Manio's Quadrilateral. The eclipse to solar year triangle can be constructed using the diagonal of three squares just as the soli-lunar triangle can be constructed using the diagonal of four squares, implying a geometrical simplicity behind astronomical time. This construction can be seen within the Quadrilateral if three eclipse years, in day-inches, are taken as the baseline from the sun gate and three solar years are raised upward. This suggests how the type B–flattened circle geometries were actually built using three squares, which then became another portable technology—like the four squares geometry— for Megalithic circle builders.

to have its eclipse and nodal behavior tied to the Trigon period, and thus the two outermost, visible planets. This gives the terrestrial precessional cycle a highly accurate synchronization to the moon, Jupiter, and Saturn; and these are the most gravitationally influential forces upon precession apart from the sun and hence the solar year. It is also true that Saturn has a synodic period of 16/15 lunar years, the same as the Trigon period's relation to the lunar nodal period. Jupiter's synodic period is also synchronous with the lunar year, but in a whole tone of 9/8 (with a unit of 3/2 lunar months; half and whole tones are explained in the following chapter), whereas there are 17 Jupiter synods in the moon's nodal period plus an additional 18.618 tropical days. At Carnac, the eastern cromlech displays this ratio of 15 to 16 between its forming circle and the outer, egg-shaped perimeter, indicating that this ratio between the nodal period and the Trigon was quantified through the megalithic astronomical work of the famous Le Menec Alignments.

As an aside: the precessional cycle of 25,740 can be factorized in an interesting way as 99 times 260 years. The Tzolkin Mayan calendar is 260 days long and if multiplied by the 365-day Haab, a Mayan solar calendar, would generate a full cycle between these calendars that is 260 practical years long. The 99 periods of 260 years then point to 25,740 years, whereas 100 such periods would equal the familiar 26,000 years often used as a first approximation of the precessional period. There is a reference in Islam

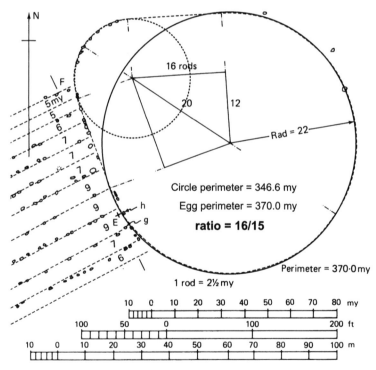

Figure 2.15. The eastern cromlech of the Le Menec alignments displays the 15 to 16 ratio between the moon's nodal period and the Trigon period of Jupiter-Saturn conjunctions. The implication is that this link between the moon and the Trigon could have been identified, the Trigon then being the traditional subdivisor of the precessional cycle of 25,740 years.

to the 99 names of Allah, the hundredth of which is unknowable, and this would be a typical way of encrypting a sacred fact in the biblical/Semitic tradition, as with the biblical allusions to the square root of two and doubling the volume of a cube, which I discuss in the next chapter.

NUMERICAL SYMBOLS OF COSMIC TIME

Ratios manifesting as numbers and geometries emanated from the Megalithic study of cosmic cycles, and this generated a new class of number symbolism that contrasted with the metaphoric symbols developed by the earlier storytelling age. These numerical symbols were related to the characterization of planetary gods, and they extended beyond the simpler categories of twofold, threefold, and fourfold divisions already obvious in the sky—found in the Northern and Southern Hemispheres; the Equatorial, ecliptic, and galactic frames; and the four gates of the sun throughout the year. Seven already represents the number of planets (5), including sun and moon (2), and the two extra nodal points in which eclipses occur, to make 9 moving factors.

The Megalithic activity expanded the range of characteristic numbers that were symbolic for the bodies that generated them, using the day-count triangles they built. These first became significant through astronomy but then became a basis for sacred symbolism involving number. This became our sacred, religious symbolism but then was obscure, rather than intelligible, in its codification of the cosmic facts.

Five for Venus, the Goddess

Venus is the brightest object in the sky besides the sun and moon. It is at its brightest when it approaches and departs from Earth in a cycle that takes 584 days. The solar year, in practical terms, is 365 days, and the common unit of time between these two periods is an accurate 73 days, of which there are 5 in the practical year and 8 in the Venus synodic period. As a consequence, there are 8 Earth years in 5 Venus periods, which means that the manifestations of Venus, such as an evening star or as a morning star, appear again within the zodiac in a position moved on from the last by 2/5 of the zodiac. In 8 years, a pentangle is described by, for example, the morning star in the heavens, and therefore Venus is most strongly associated with the number 5 and with the 5-sided star.

At the root of 5 lies the golden mean, a touchstone ratio for use in architecture, most notably that found in ancient temples, especially Greek or Roman, or in modern styles that borrow these classical styles. Venus was said to be the cosmic generative principle cut away from the sky god by Saturn, the god of time.

Seven for Saturn, the God of Time

The synodic period of Saturn is 378 days, which divides by 7, making the 7-day week commensurate with Saturn's synod of 54 weeks. The 52 weeks familiar as our year is not exact, because $52 \times 7 = 364$ days, and this causes the daily movement of dates, such as birthdays, every year. Yet this Saturnian year of 364 days was widely observed in the pagan world, working with exactly 52 weeks or 13 months of 28 days, and was associated with a matriarchy that elevated men as kings who ruled for a year and a day—that is, a solar year of 365 days. It was thought that these kings eventually took over, to create patriarchal societies with the Olympic sky gods who have come to dominate the West. The leader of the Olympiads was Jupiter, who had displaced Saturn and the original Titans that operated the mill.

Twelve and 60 for Jupiter, the King of the Gods

Jupiter is titled King of the Gods, and he stands for the perfect numbers, such as 12, based on 2 and 3. Twelve times 5 is 60, and 6 times 60 comprises the perfect regularity of 360 degrees in a great circle. Though there are 12 lunar months within the solar year, forming the lunar year, the synodic period of Jupiter is 9/8 greater than this, a perfect whole tone (see next chapter). Jupiter therefore owns the moon, because the two bodies are in resonant interlock.[5]

If the moon is a key part of the mill, then Jupiter holds the handle with his whole tone. Yet Saturn is to be found behind him, since Saturn has a semitone relationship to the moon, and the Jupiter-Saturn Trigon period is commensurate with the precessional cycle. Jupiter also sweeps out 1/12 of the zodiac in 361 days, 361 being the square of 19 leading to relations to the longer Saros and Metonic periods.

Nineteen for the Moon, the God of Measure
The whole of the moon's essential behavior is contained in the Metonic cycle of 19 years, and the eclipse phenomena operates between 18 and 19, as the Saros period of just over 18 years (for the repetition of similar eclipses) and the lunar nodal period of 18.618 years. Nineteen bluestones at the center of Stonehenge, therefore, pointed to lunar symbolism, especially since the bluestones themselves, partly quartz, have a lunar designation.

It is extremely unlikely that the sun, moon, and stars coincide in such a short cycle and exactly on an anniversary or year end. The Saros cycle of eclipses consists of 19 eclipse years and 223 lunar months. Nineteen lunar years contain 228 lunar months (5 more months) and 19 solar years, the Metonic cycle, 235 months (7 more again) making 12 months, or one lunar year, between the length of the Saros and Metonic cycles.

The Music of the Spheres
Numerical astronomy reveals ratios, and when these were found to be particular, whole number ratios, such as 9/8 (whole tone) or 16/15 (half tone or semitone), then the planetary spheres came to be seen as having some musical content. This idea was taken to the extreme in the Age of Aries, when massive tone matrices made up of only 2, 3, and 5 were built (see chapter 3). This established a new playground we call arithmetic and abstract analytic mathematics and storytelling based upon pure harmonic facts.

A MOON THAT CREATED EARTH

MORE THAN 4.5 BILLION years ago, the inner solar system was a jumble of would-be planets and planetoids. It is thought that Earth shared its orbital zone with at least one competitor, about the size of Mars, similarly composed of a heavy metal core and outer mantle. Both planets would have mopped up smaller bodies, but eventually the two collided with each other.

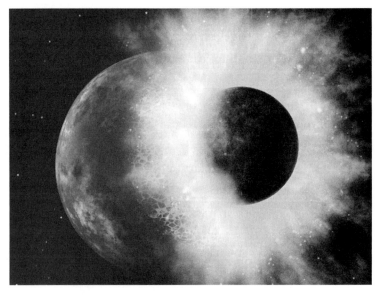

The collision of Earth and Thea; courtesy NASA/JPL-Caltech

In a huge explosion, Earth was severely impacted, and the energy release caused vast amounts of both planets' surface rocks to be vaporized or projected into space. This resulted in a ring around Earth that quite rapidly consolidated into a single body, which soon cooled and aggregated into the Moon that orbited Earth every 20 days, a mere 2,700 kilometers (1,690 miles) above Earth's surface. The planet that struck Earth has been called Thea, after the goddess that gave birth to the moon, called Selena in Greek myth. Meanwhile, the metallic core of Thea was not absorbed by Earth's core. Instead, significant metal deposits were embedded in the surface layers of Earth—which gave Earth's crust a rich set of workable ore deposits named the "wedding ring." This was significant to the later metalworking ages.

Such a massive satellite traveling around Earth caused the whole surface of the planet to deform gravitationally below the moon, but Earth was then rotating every 6 hours, so that this bulge would always be dragged ahead of the moon. Just as with tides today, but much more strongly, Earth's rotation transferred energy to the moon, causing the satellite to accelerate and take an ever-higher orbit. This, however, did not occur before the moon had kneaded all the surface rocks in the Earth's tropics. Subsequently, this type of lunar influence continued in an unusual way.

Around 4 billion years ago, the orbits of Jupiter and Saturn aligned to create a slingshot for solar system bodies that had not yet been incorporated into planets. This Late Great Bombardment proceeded to strike the moon rather than Earth, and this protective role of the moon probably saved Earth from damage to its nascent resources, such as the water present on its surface. The recognizable face of the moon was largely created from this bombardment, as the craters and seas of molten basalt we see today.

By three billion years ago, the moon had been accelerated to an orbital distance of 320,000 kilometers (200,000 miles), which meant that its tidal effects were no longer strong on the earth's crust but instead, the seas of that period experienced massive tides that were hundreds of meters high. These must have been like continuous tsunamis that raced around the globe. Moreover, due to the still-great rotational speed of Earth, these tides occurred many times a day, and in addition, the early atmosphere was whipped up by the Coriolis effect, which created continuous, hurricane-speed winds. The extreme ocean tides, causing massive erosion and mineralization of the seas, formed a number of chemical scenarios that might have facilitated the creation of life in the form of primitive replicating molecules.

Even today, volcanoes and earthquakes are thought to trigger eruptions and the release of seismic energy, built up in Earth's unique tectonic plates. These plates themselves could be an artifact, in part, of the moon's kneading of Earth. For instance, we can see that on Mars any plate activity ceased billions of years ago as the mantle became stuck to a solidified core—probably because of its lack of a large moon in that planet's orbit.

Though the original collision almost certainly caused the great rotation of Earth, it also created the tilt of the axis on which Earth rotates. This tilt set up on our planet the seasonal conditions that have been so important to life's varied habitats. Yet this tilt would not have been stable without the large moon. Our moon stabilizes the tilt by shielding Earth from the small, chaotic forces our planet experiences due to the other planets. Mars is particularly vulnerable to these forces, and its tilt varies, over millions of years, by about 30 degrees. The moon, by adding a large, systematic component to the precessional forces, prevents planetary chaotic resonances from affecting the Earth.

The effect of the seasons, maintained by the moon, is joined by the extra tidal effect the moon has on the seas and oceans. These tides create an extensive area of a very important habitat: the tidal ranges found on our coastlines. These areas are very diverse biologically, and they have led to evolutionary changes as significant as the adaptation of marine animals into land animals.

In summary, life on Earth would not be possible without the moon and the special itinerary of its genesis and the gradual arrival to the conditions we find today. Perhaps this role of the moon has lead to the general recognition that life on Earth could not have evolved without a special collision that occurred at the exact distance from the sun that was capable of supporting such life. The precessional mechanism appears necessary for a planet to support distinct and stable land habitats for life, and therefore, our large moon was an essential accessory for our form of existence.

CHAPTER 3

GOD AS HARMONY IN ARIES

THE LITTLE-KNOWN ACHIEVEMENTS and discoveries of the Megalithic period emerged through steps involving counting and the geometrical properties of circles and triangles. Once this geometry had arisen, based on counted lengths, it developed a metrological toolkit in which types of foot measurement could perform tasks such as obtaining an integer perimeter from an integer radius by building near-Pi ratios such as 22/7. By the end of the Megalithic age, it was obvious to those that understood it that the cosmic world was an artifact built upon numbers—but what were numbers?

The answer to this question leads down a different road than that of megalith building. The constructing of temples employed sacred numbers as a norm for religions in the Age of Aries, but though these temples were large, and in some cases aligned to the heavens, these temples did not focus outward: they sought to create religious spaces for the gods. Neither did these interior spaces marshal the dead for an afterlife, as burial mounds had.

This was a world torn asunder by Iron Age conflict, with gods having the power to bestow victory or defeat. Civilization, unified by the trade but riven by conflict, had been established in the Bronze Age—and only trade could make fortunes and create empires, dynasties, and great cities. The emerging numeric and arithmetic skills became as important in the measurement of trade goods and debts as they were to the new abstract abilities of astronomers who, instead of constructing megaliths and building proportional calculators, made tables of measurements, developed numerical shortcuts, and calculated results. The role of numbers as a tool had changed after the Megalithic period.

What happened to the Megalithic knowledge in the recently inaugurated practice of writing history? There are few written traces of most of the results presented in chapter 2, except that a new type of quasi-secret activity began—an activity conducted by priestly scribes and focused on numerical harmony. Evidence of this has recently emerged, showing most of the surviving literature from the Age of Aries to be encoded using a very specialized knowledge of musical harmony. We also find that the metrology of the Megalithic period expanded into a larger set of standard measures, especially suited to representing musical tones and hence harmonic concepts.

In the Age of Pisces, the resulting literature—such as the Rig Veda, Homer's epics, the Bible, and other texts, especially the stories of Semitic religions—came to affect intellectual and religious life. These stories, told as factual (hence "literal") but largely based upon sacred numeric ideas about harmony, were a major intervention in human life. Specialists in many different regions carried out this intervention, but they were probably descendants of the specialists who, in the Age of Taurus, had discovered the secrets of the cosmic world and were now priestly scribes.

The mutation of megalithic metrology into the rational world of harmony was like a move from applied mathematics to pure mathematics: the subject became the generative power of numbers rather than their practical application, and cosmic time periods were no longer the generators of significant ratios.

FROM METROLOGY TO MUSIC

The first realization about harmony was the need to establish the mathematical constraints on what are harmonious intervals, an interval being the ratio between two notes. Early music had discovered that strings vibrated according to their length and that the harmonious intervals between notes, seen as a ratio between string lengths, were few and definite to produce harmony to the ear. The Indians suggested the universe was made of sound and therefore embodied number harmonies. Such ideas establish a link between cosmology, metrology, and the harmony found in the lengths of strings that generate sound. Metrologists were well equipped to study the mathematics of music (it is simpler than astronomy) and to take the early steps in acoustical theory. This exploration was conducted by using small musical instruments such as the monochord—a single string running over a sounding board whose vibrating length could be altered. Intervals could then be measured in terms of ratios between string lengths.

We know of a broad range of standard length measures, named after the places where they were first discovered in practical use or where they were clearly involved in the construction of local monuments. The royal cubit of 12/7 feet is, for example, based upon a foot of 8/7 feet but it is 3/2 larger. This royal unit of length is considered Egyptian and can be found throughout the dynastic Egyptian monuments. It was widely used in that region, where gods and god-kings were shown pictorially as 8/7 taller than ordinary men. The ratio of 3/2 between string lengths generates an interval called the perfect fifth, which is the most harmonious interval after that of the octave ratio of 2/1.

In fact, the most unusual thing about historic measurements is that they all have a single root since they are all fractions, or ratios, of an English foot, and these fractions are largely composed of harmonic intervals such as 9/8, 10/9, 16/15, 21/20, 25/24, and 36/35, all of which are found in Plato's defining works from the ancient school of musical harmony. Plato wrote about the world of musical harmony in an explicit way, whereas the norm at his time was to create the harmonically encoded stories found in the scriptures and epics of the period. The underlying structure of these stories could be understood only by those initiated into these harmonic mysteries. We must be grateful to Ernest McClain who has rediscovered, in recent years, Plato's harmonic science and hence revealed what lay behind the harmonic epics and scriptures of this Age of Aries.

Long before Plato, however, metrology appears to have been an early tool used now to explore a universe of harmony rather than the time periods of astronomy, as is clear in the harmonic fractions built into the historical measures of length. Otherwise, why should the different types of feet encompass all of the harmonic ratios?

In air sound moves in a series of compressions and rarefactions, and the distance between successive compressions is called the wavelength. The higher the pitch of the sound, or what is called the frequency, the shorter its wavelength. In a stretched string, notes are made by the string's oscillation. Instead of the energy traveling from one place to another as in sound waves moving through the air, in a stretched string the energy is contained in what is called a *standing wave.* Its basic note has a wavelength of the length of the string. If this length is halved, then the frequency or pitch is doubled. The interval between these two notes is called an *octave,* and its two notes are often referred to as "low do" and "high do." The number 2 creates this first harmonic fact.

In changing from one note to another, we move up or down in string length. The resulting change in pitch is called an interval. In classical "pure" tones, rather than in the compromises of modern tuning, each interval is a numerical transformation represented by a fraction in which only the numbers 2, 3, and 5 are present as the numerator and denominator, respectively. It is a miracle that just these three *prime numbers* can create all the intervals required to fill an octave's doubling in frequency so as to provide the world of musical harmony.

Within the span of octave doubling, the next highest number after 2—that is 3—combines with 2 to generate the next two most harmonious intervals: the fifth (3/2) and the fourth (4/3), plus the whole tone 9/8, which is the interval between them. An ascending fifth from low do is also a descending fourth from high do. The next prime number is 5, which leads to two new intervals, another whole tone (10/9), and the semitone 16/15.

When intervals are "added," they are in fact multiplied, because they exist within a logarithmic world (base 2 because of octave doubling). For every note there are many octaves above and below it, each of which is a doubling or halving yet each sounding the same only higher or lower. Within each octave, the same harmonic intervals are always found because each octave is always just a doubling, independent of what note is actually being doubled. One can therefore call the octave and the possible intervals within it an *invariant* property of the universe, and this ancient science of musical harmony, when encoded into the scriptural and epic writings, is *The Myth of Invariance* referred to in the title of McClain's widely published book. (As we have seen in previous chapters, astronomy and ring composition were the myth-generating invariants from previous Ages.)

Adding the interval 10/9 to the other whole tone interval (9/8) generates a major fifth, because these two tones reduce to 5/4 *when multiplied* (the nines canceling leaving 10/8, which is 5/4). Harmonious intervals of this kind determine the form of what we call a scale, such as our diatonic scale with its familiar notes of low do, re, mi, fa, sol, la, ti, and high do (see figure 3.1 on page 74). This form is invariant, or constant, and exists within the world of number itself though experienced directly by the ear and, in music, by the mind. These intervals can be arranged in six different sequences with regard to placement of the semitones (16/15), leading to the familiar modal music of different scales. Each scale is made up of the five whole tones and two semitones that always add up to octave doubling. (A weak cosmic parallel to this is 5 visible planets and 2 luminaries—the sun and moon—which, when added, equal 7.)

Pure tone harmony is limited in that it must provide intervals that guarantee exactly matching octave doubling. As we have said, this harmony is built using only the first three prime numbers—2, 3, and 5—and it is a fundamental tenet of musical harmony confirmed by the ear yet wholly numerical. Yet if string lengths are defined using metrology, even if the metrological system has all the requisite ratios ready-made, the task is challenging. Any given length defining a note must be able to be scaled up or down to allow others to be harmonious to it. Fortunately, in going back to thinking about lengths with respect to a right-angled triangle, the triangle's slope angle is such that any length placed on its base can be projected upward onto the hypotenuse to scale up that length by the ratio of the triangle's longest sides. In other words, the slope angle defines a ratio that can multiply any string length and establish a new string length so as to achieve any required interval. Division by the same ratio can also be achieved by reducing a length, by projecting downward from its hypotenuse to its base, so as to increase the frequency of sound produced by a string.

The reason why our historical measures are harmonically related to the English foot becomes clear. Metrology was used to construct the string lengths within a study of harmony; using the same number of two different types of foot defines a length for the hypotenuse and base for a harmonic ratio, seen as a triangle. Originally, making harmonic measures would have involved the creation of a suitable triangle having the necessary ratio in its side lengths, measured in English feet. Once a range of such harmonic interval measures was constructed, there was no longer any need to construct triangles every time. Instead, an English measurement ruler can be used to measure a string's length and that same number of units using a different foot would achieve the required tone's string length. It would have been possible to build an early fretted instrument; the distance between frets marking the differing string lengths for the same string, when pressed down to the fret.

The implication is that our historic measures were survivors of metrological kits used for harmonic work and that only some of these, such as the English inch and foot, were the earliest measures used originally for megalithic astronomy. This is confirmed by the nature of the microvariations found in historic measures that have a relation to Pi and also of our modern tuning system called equal temperament. (See An Equal-Tempered Zodiac, page 87.)

THE JOURNEY INTO HARMONIC NUMBERS

When numbers came to be notated in a written form they provided an alternative to using metrology. It was discovered that some of the whole numbers, between a starting number and double that number, were separated by the harmonic intervals found within an octave. This enabled the study of harmony using strings with whole number lengths using any unit of measure.

For example, the fifth, fourth, and whole tone intervals (involving the number 3) could be generated between the number 6 and 12 (= 2 × 6).

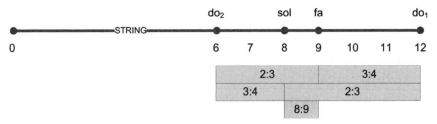

Figure 3.1. The perfect harmonic intervals provided by the integer numbers between 6 and 12

Music within a Conceptual Month
Plato used rules for inviting friends to a wedding or training guardians of a musical city to allude to the pentatonic scale formed from sets of numbers that are able to define pentatonic tuning in this way, number sets starting with 24 or 30 and therefore ending with 48 or 60. (See figure 3.2.)

It is clear that intervals involving the prime number 5 are needed to create pentatonic tuning using these numbers. However, in the scale shown above there is limited capacity available for melodic music as the intervals do not include any semitones between the pentatonic notes.

Music within a Conceptual Year
To achieve what is called just tuning, with more potential for making music, a larger starting number is required. This will enable more inter-

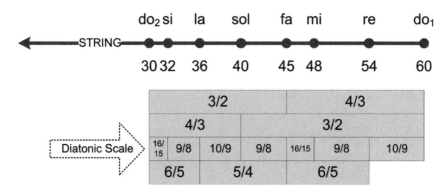

Figure 3.2. The perfect harmonic intervals provided by the integer numbers between 30 and 60. The word diatonic can be understood here as the scale of seven notes within an octave that contains two semitone intervals—a pattern seen in the white notes of a piano that, in two places, have no black notes in between.

esting intervals since the number of available whole numbers within the octave range has increased, and some of these will land on notes (as numbers) that are not available in a smaller octave range. The smallest number range to achieve this is 360 to 720. (See figure 3.3.)

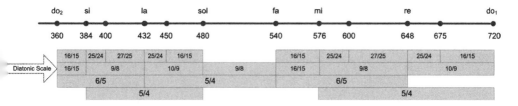

Figure 3.3. The perfect harmonic intervals provided by the integer numbers between 360 and 720

McClain says, "At this level of development, the scale 'bent round into a circle' as required in *Timaeus* (36c), can function as a zodiacal symbol. . . . Since these tones are distributed within the octave in perfect inverse symmetry, the same set of integers serves both the rising and falling scales: they could be applied to the tone circle in either direction. The 360 arithmetical subdivisions within the octave 1:2 = 360:720 correlate with the 'idealised year of 360 days in some ancient calendars.'"[1] (See figure 3.4.)

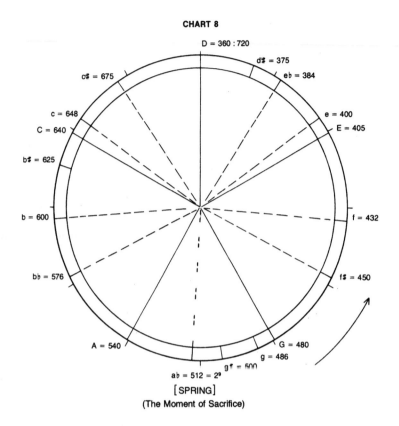

CHART 8

Figure 3.4. Chart 8 of McClain's *The Myth of Invariance* showing the tones of 360:720 "bent round into a circle" (in a logarithmic sense) to approach a zodiacal symbol of twelve equal intervals, if a note directly below D is added. This means that twelve semitones are now available and that more than a major scale can be employed, as is found within modal music with six different permutations of the two semitones leaving five whole tones. The leading whole tone of 9/8 can be achieved using 405, shown as E, only because there are so many more integers available within the larger octave range of 360:720.

As larger integer sets are generated there is an increase in the number of harmonic intervals, and these tunings offer more sophisticated musical scales. Different starting numbers offer different sets of intervals as tuning schemes, and these appear automatically from the choice of a given starting number. While every integer ratio marks a possible interval, most of these are not harmonious within the whole octave.

The rejected numbers and notes within these harmonic intervals can be seen as *sins* in that they "miss the mark" set by harmony, and from this, harmony naturally emerges as an implicit value system. Such a system of values, set by an unknowable God who created the universe through numbers, is an extension of the Megalithic project concerning numbers and the cosmos. The elites of the Megalithic period doubtless became entangled in this new Age's preoccupation with harmonic numbers, where such calculations led the way to founding new religious doctrines based on the bedrock of harmonic number and the vision of a divine harmony.

In the minds of number technicians the new harmonic realities became self-evident, but to a population of Iron Age traders, there were more pressing, everyday concerns to keep them occupied, such as trade. The two focuses coincided, and while trade had revolutionized the ancient world, popular ideas of the sacred were still embodied in the people's stories. The power of stories is that within each successful story lies an implicit worldview and, within that, a "big issue" or moral quandary that had to be resolved. The world of harmony appeared to generate tuning systems that showed a fight between good intervals and bad intervals, where good meant the greatest harmony and bad was represented by intervals that interfered with the orderly flow of the octave and had to be overcome or redeemed. This was an implicit worldview for a story within harmony but, to reach the public ear, the narrative could not reveal itself as based on numbers.

A way forward emerged in which different cultural groups would create their own stories based on characters and scenarios drawn from different tuning systems. The starting numbers and hence tuning systems inspiring the Bible were different from those used by Homer (13,500 lines in the original *Iliad*), while those of the Hindus were extremely ambitious in their use of very large numbers, such as their number for world Ages (units of 432,000 "years"). However, as the numbers increased in size, it was hard to see the significance of what was really happening between them; this research program, based on finding the tunings possible using starting numbers for the octave, rapidly exceeded the acoustic world of music as experienced through musical instruments. An intellectual "space race" ensued—carried on for millennia and leaving our oldest books as its harmonically coded documents.

CLIMBING GOD'S MOUNTAIN

The number 2 is not the cause of these new worlds of intervals, as starting numbers increased, because any number doubled only forms an octave interval. How could one tell which starting numbers were good to try? The purest harmonic tones belong to integers that differ only by 2:3:4:5:6, and if 2 and 4 are removed from this list then the prime numbers 3 and 5 are the creators of harmonic intervals within octaves. Higher starting numbers unlock new interval ratios by using powers of 3 and 5 to make up the new whole number tones.

According to McClain, the musical number theorists created a grid of all these powers of 3 and 5 in which the columns were increasing powers of 3 and the rows upward were increasing powers of 5. (See figure 3.5.) A general-purpose grid was calculated covering ever higher powers of 3 and 5 and their products. It was a work that needed to be done only once in each scribal center, with the results saved in the number notation of that place and made available to fellow workers.

Step 1: Limit the Powers of 3 and 5

Using the grid of the powers of 3 and 5: what about the limiting number of 60 used for the pentatonic scale? The process is shown in figure 3.6 on page 81, and it yields the same set of numbers but this time automatically—that is, using only the power grid of 3 and 5. (The grid is skewed right to show the two types of third, major and minor, as triangular relations within the grid—a technique that also proved visually evocative.)

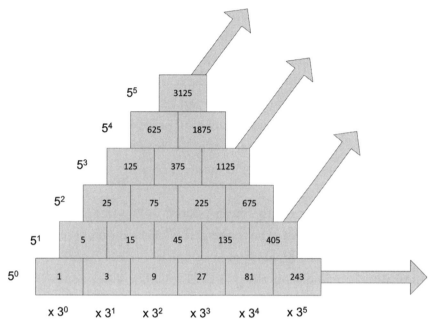

Figure 3.5. The powers of 3 run left to right, and the powers of 5 run from bottom to top, to form the essential ingredients for any intervals within an octave. The building blocks are each the product of 3^p (to the power p) and 5^q (to the power q), and, because any number to the power 0 equals 1, the grid starts with $3^0 \times 5^0 = 1$ on the bottom left, as shown, for most of the ancient schools of harmony.

Step 2: To the Limit, Double the Numbers that Remain

Once the active ingredients for any possible harmonic interval, the powers of 3 and 5, have been limited (here, to 60, giving 45 as the highest number), how can an octave be found within the grid? The number 2, required for octave doubling, is as yet wholly absent within these powers of 3 and 5. The next key concept (see figure 3.6) was to again enforce the limiting number, but this time for every number. Each number can be multiplied by 2 the number of times that prevents it from exceeding the chosen limit (in this case 60), and then all the integers in the grid will belong only to a single octave range for that limiting number as high do. That is, all the numbers are doubled until another doubling would exceed 60, so 30 would qualify for doubling but no number higher than 30. This was probably the world's first computational algorithm.

The shape of McClain's yantras often resembles a mountain when the limiting number reached at the top is a pure power of 5; because each layer below employs a lower power of 5, and hence requires additional powers of 3 to achieve the limiting number, the base of the yantra is widened into the form of a "mountain." The simplest yantra of all, shown in figure 3.6, expresses all the combinations of $3^p \times 5^q = 60$, for all values of p and q. The yantras tell of the type of gods evolved in the post-Megalithic period—gods with numeric properties that belonged to the harmonic project.

In this yantra for 60, the mean position is $2^2 \times 3 \times 5 = 60$ and is associated with Anu-An, who is the god of that yantra and represents $do_1{:}do_2$ as 30:60. The highest power of 5 is its square, $2 \times 5^2 = 50$, traditionally the number for the god Bel-Enlil. The lesser power in the middle row is associated with the god Ea-Enki whose number is 40 and whose position is in the middle row, left, with the formula $2^3 \times 5 = 40$.

The methodology of working within numerical limits is defining a religious pantheon of numerical gods. From this view there flowed many innovations, such as using 60 as a harmonic base for the Babylonian number system, the use of 360 degrees, and a system for dealing with the arithmetic of doubling and the ability to generate ever-larger yantras.

The God on the Mountain diagram on page 81 produces the lowest set of numbers that can generate a useful scale. Yet it also introduces a new meaning in which the relativity of all tones can be known from the fact that

- traveling right generates a rising fifth (i.e., multiply by 3/2)
- traveling upward and left to right generates minor thirds (multiply by 6/5)
- traveling upward and right to left generates major thirds (multiply by 5/4)
- reversing these directions generates reciprocal results
- string lengths are the inverse of string frequencies

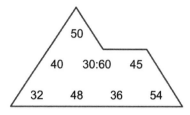

Step 3: Yantra for N < 30
"God on the Mountain"
The Diatonic Scale of
[30 32 36 40 45 48 (50) 54 60]

Step 2: Double to N > 30
(Powers of 2)

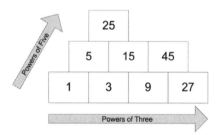

Step 1: Building Blocks for N < 60
(Powers of 3 and 5)

Figure 3.6. McClain's rediscovered method of investigating the pure tone intervals within any integer range, here presenting the range 30 to 60. In step one (starting from the bottom of the image and working your way up to the "top of the mountain") the reader will see the top brick is five squared (25) and the middle brick is three times five (15), showing the blending of powers of three and five within the wall. All these numbers are doubled until a further doubling would exceed the limit of 60. The full scale of intervals is then available as integers, and they are guaranteed to express harmonic intervals between them. The design is similar to those yantras still drawn in India (e.g., the Sri Yantra; yantra is the Sanskrit word for "instrument" or "machine"), and the diatonic scale was the one seen previously for 30:60.

The God on the Mountain outline is evocative, as are the directionality and tonal characteristics that generate a tonal topography for storytelling. This was seized upon in order to write scriptures. New types of scripture from the Sumerian experience onward appear to use the tonal yantra as a template for the codification of spiritual myth making. This continued through the Bible's Old Testament (Hebrew Scriptures) to the end of the New Testament (Christian Scriptures) and in the Book of Revelation. The New Testament was heavily influenced by the numerical knowledge of the harmonic schools, knowledge that Plato had made somewhat public in explicit form in his allegories about different cities: the Greeks were everywhere in the eastern Mediterranean at the time of Christ, and the harmonic initiates were amongst them. Different scriptures and epics introduced different limiting numbers and then portrayed the tonal drama unique to each, as an inspiration for the characters and dramas involved.

The power of the yantra system was that a whole set of tuning issues could be stated as this single limiting number, mentioned once or twice within a text, and anyone knowing this system could reconstruct the yantra and read the remaining text and its characterization of people, situations, and other numerical clues—a process of decoding the texts repeated by McClain in our time. The technique preserved harmonic knowledge as sacred stories in the Age of Aries.

THE HARMONIC LIMIT FOR PRECESSION

The precession cycle itself can form a yantra, which McClain provides in *The Myth of Invariance*.[2] The Platonic Year is a precessional period chosen as being numerically suited, in length of years, to harmonic factorization (being factored only by the numbers 2 and 3). McClain has investigated the limiting number of 25,920 and created a yantra for it that reveals the harmonic intervals between a low do of 12,960 and a hi do of 25,920. (See figure 3.7.)

In figure 3.7, apart from the seven capitalized notes of the diatonic octave, available through the 30:60 octave, the elements constitute a system for the limited key changing possible within a just tuning system. In this octave of precession, the only ratios lying between its notes are just four semitones, and these prove significant to later ideas about the moon and precession and the *syntonic comma* of 81/80.

- the Pythagorean semitone of 256/243, which corresponds to the ratio of the Metonic period to the Saros period
- the diatonic semitone of 16/15, found between the Trigon period

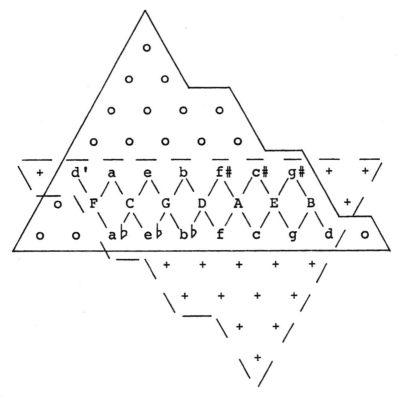

Yantra for the Precessional Cycle of 25,920 Years

Figure 3.7. The tonal yantra resulting from a limiting number (do₂) for the precessional period taken to be 25,920 years long (from McClain's *The Myth of Invariance*). The mountain shape is inverted as a dotted outline in larger yantras such as this, relative to the "lynch pin" of D, the do₁:do₂ octave: notes contained within both mountain outlines are then relevant to both falling and rising octaves, representing reciprocal "twin" notes that are symmetrical to left and right of the tone-mandala (figure 3.8) for the yantra.

(that divides into the precessional period) and the moon's nodal period (and also Saturn's synodic period, relative to the lunar year)
- the chromatic semitone of 25/24
- the syntonic comma of 81/80, which can transpose between 5-tone (pentatonic) and just tuning and which releases the diatonic scale from some of its limitations, for example 256/243 is 81/80 shorter than 16/15

- the ratio 27/25, achieved by combining 16/15 and 81/80, which is found in the ratio of lunar month to lunar orbit

The semitones that appear with this yantra are therefore a synthesis of the tuning systems for 6:12, 30:60, and 360:720—the systems of the conceptual zodiac, month, and year within the harmonic world. These tuning systems are present in the limiting number of precession by virtue of the fact that, in an octave that ends with a high do numbering 25,920 there is a rich combination of just these possible intervals.

CHART 22

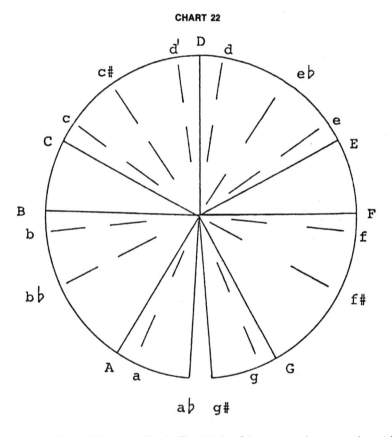

Figure 3.8. Chart 22 of McClain's *The Myth of Invariance* showing a logarithmic mandala, base 2. Precession has a duration of about 25,920 solar years, which can be viewed as a frequency. This limiting number is then generating a set of intervals for the Earth. Precession, through this number, would contain definite intervals within it, with the years in a precessional cycle as high do and half of this a low do.

Chapter 6 gives further insight, based on the tones 256/243 and 27/25, which accurately define the ratios of the Saros cycle to the Metonic cycle and lunar orbital period to lunar month, discussed by the end of chapter 2. The ratio 81/80 can also be correlated to the claim that the moon affects life on Earth, which implies that the earth-moon-sun system is organized to generate tones that extend Gurdjieff's proposal of a cosmic tuning system. (See chapter 5.)

THE ADVENT OF A SINGLE GOD

Hundreds of years before Christianity and Islam arose, the Bible's Old Testament was written using harmonic ideas such as purity, pure order, and an invisible and jealous God. But the general population had neither the time nor the inclination to worship such a God directly, because the cosmic mystery of Earth had been the substratum of the Megalithic period, when the gods were clearly in the sky. However, civilization was becoming an ideal of social order beneath god and king or emperor, according to ordering principles of rulership. Those who thought differently became pagans for the Romans who understood the Platonic ideals of state organization based upon harmonic concepts.

The work on harmony had transmitted to the mind something more powerful than beliefs, something that organized thought itself for the rational mind. This mind's cultured education became the discriminator against the uneducated, whose identification with the land, ancestors, and mysteries disordered the objectives of the state or empire, in which controls were necessary. A harmonic parcel bomb had been loaded into scripture by adepts and would explode to create our history and the great Western religions.

On one level, the topography of a tonal matrix seems an unpromising start for scriptural content, though in retrospect it has been effective. Therefore, we must consider the possibility that these gods of harmony were indeed an effectual and powerful matrix for the mind, perhaps none other that the newly rational mind of humans, which was seeking its own order. This was not the demiurgic mind, the numeric intelligence ordering the local cosmos and now considered regressive. The rejection of the many gods, clearly based on the planetary reality, was a rejection of the earlier mind's interest in the cosmic machine as a source of knowledge. The new mind machine could and would generate ever more information and knowledge under a god of pure symmetry and mental relativism; and categories of relevance emerged, all of these being tonal concepts.

In breaking free from the cosmic order by the end of the Age of Aries, the mind had become conscious of itself and was given new powers to think outside the box of cosmic actuality. Our modern ideal of free speech would horrify the ancient priests, whose purpose in articulation was the transmission of objective knowledge. These mores existed as late as the last great Welsh bards, whose objective poetic formulations were said to carry the power of life or death. In the New World such magical power was regressive, because all was to be held together by the mind rather than by the magical order of the natural world. This schism is fundamental to this day, and it has been continued by our modern science that explores the natural order while still not recognizing the Megalithic achievements. Within the last six thousand years, the probable sequence was:

- the mind itself came to Earth from the gods, through the numbers in the sky
- the priests and prophets revealed the one god of the mind's harmony through the musical harmony that forms the heart of the number field.

In this way, the mind realized itself first through the sky and then through harmonic number field and became conscious of itself. By the time of the rise of Athens, humans within classical culture had started to become individual, a process called in psychology *individuation*.

AN EQUAL-TEMPERED ZODIAC

The form of musical scale we use today is the equal-tempered scale. Its capabilities express well the new mind's freedom of movement, allowing us to change key and to thereby play between alternative keys (different octaves, separated by a musical interval). This equal-tempered possibility was alluded to in ancient harmonic theory and was approximated as the constitution of one of Plato's city-states called Magnesia.

The equal-tempered scale divides the whole of the octave into twelve *equally increasing* semitones, and in the outer world, this scale is echoed in the division of the sun's path within the year, as it is seen as passing through twelve equal sectors called the zodiac. The Vedas seem to refer to it as:

> *Twelve spokes, one wheel, navels three.*
> *Who can comprehend this?*
> *On it are placed together three hundred and sixty like*
>   *pegs;*
> *They shake not in the least.*[3]

To divide an octave into twelve equal semitones, we must find the twelfth root of 2, the octave doubling. We can use a calculator to reach this value—1.059—but such calculators belong to very recent times. Achieving this value is not the only challenge, as one must develop the full scale of these twelve notes, and these values are not available through the numerical yantras for any limiting number.

This twelfth root of 2 could only have been generated, if at all, through the Megalithic geometry of the $N : N + 1$ triangle, because these triangles can simulate growth based upon any logarithmic base, and the logarithm (base 2) of the twelfth root of two is numerically 1/12. It is this interval that, in an equal-tempered "zodiac" of tones, can form a tone circle divided into a "wheel" with "twelve spokes," each equally spaced. (See figure 3.9.)

CHART 2

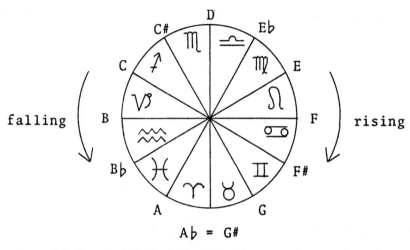

Ab = G#

Figure 3.9. Chart 2 of McClain's *The Myth of Invariance* showing a logarithmic mandala, base 2, of the zodiac as an equal-tempered scale.

If the base of an N : N + 1 triangle is given the conceptual value of 1, then there must be a value for N such that the N + 1 hypotenuse equals the twelfth root of 2 relative to the base. Furthermore, once this triangle is constructed, then N + 1 can be arced down to the base and then raised as a perpendicular to strike an extended hypotenuse at the point (N + 1) times (N + 1)—that is, an equal-tempered whole tone of two semitones.

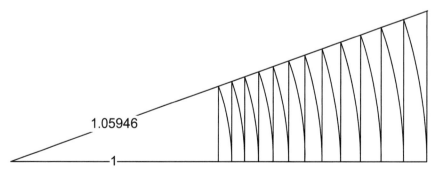

1.05946

1

Figure 3.10. An N : N + 1 right-angled triangle that can generate the twelfth root of 2, suitable for the equal-tempered scale. The pattern of frets on a guitar are identically spaced to the lengths generated by this triangle.

We can then see that the successive applications of the same procedure result in dividing the range 1:2 into twelve equally growing intervals. What we now must determine is the value of N that can give access to this capability in order to generate an equal-tempered scale.

The triangle's sides are close to 1/17 different and are therefore close to the 16/15 pure half tone but are a little more distant from the septenary half tone of 15/14. However, 15/14 (= 1.07<u>142857</u>) is different from $^{12}\sqrt{2}$ to one part in 175 (one of the metrological "tweaks" found in historical measures). If 15/14 is reduced by $(175/176)^2$, then the hypotenuse becomes 1.059287, which is within one part in 6,044 of $^{12}\sqrt{2}$. This ratio, 176/175, is a variation found in ancient measures.

Another problem of the ancient world points us to the equal-tempered semitone: How can the volume of the cubit altar of Apollo be doubled? Today, it is known that a perfect solution to this question is impossible; however, the answer can be found *in rational terms* by applying an approximation to the four equal-tempered semitones.*

By multiplying the cube's side length by 5/4, the major third, but enlarged by 126/125, one can double the volume of a cube. This is a formula found in metrology, music, and myth. McClain says, "Surprisingly, the 'cube root of 2' correction of 1/125 for 5/4 probably was known in the 3rd millennium BCE stories of Gilgamesh."[4] This same correcting ratio was known in ancient metrology by both its known adjustments $(176/175) \times (441/440) = (126/125)$; that is, 1 part in 125, and its accuracy in doubling the cubic altar is just 1 part in 16,000—well beyond the accuracy of any practical metrology.

Returning to the N : N + 1 triangle that generates the twelfth root of 2, we can ask: What is the height of its third side? It is close to 7/20. If we scale the triangle to a base of 20 units, then the height is 7 units, which gives us a very simple way to arrange such a triangle in practice: we can

---

*The twelfth root of two multiplied by itself four times equals the cube root of two, the cube of which is 2, the doubled volume. So each side of the original cube (side length = 1) must increase by the cube root of two so as to double the volume, and this is equal to four equal-tempered semitones.

construct a 20-unit base and can raise a 7-unit perpendicular. The resulting hypotenuse is $20 \times {}^{12}\sqrt{2}$. Alternatively and conceptually, if a metrological "step" of 2.5 royal feet is used for the base, $5/2 \times 8/7$ feet = $20/7$ feet, whereupon the perpendicular (third) side is $7/7 = 1$ English foot.*

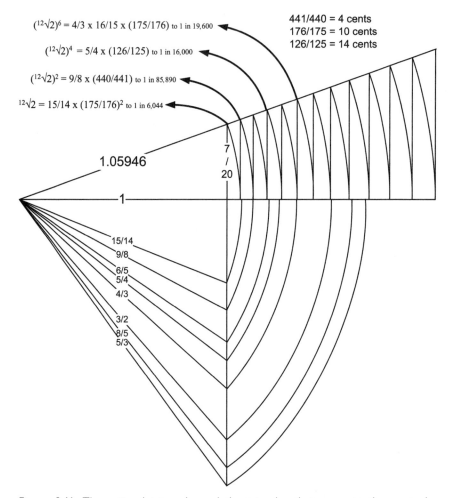

$({}^{12}\sqrt{2})^6 = 4/3 \times 16/15 \times (175/176)$ to 1 in 19,600

$({}^{12}\sqrt{2})^4 = 5/4 \times (126/125)$ to 1 in 16,000

$({}^{12}\sqrt{2})^2 = 9/8 \times (440/441)$ to 1 in 85,890

${}^{12}\sqrt{2} = 15/14 \times (175/176)^2$ to 1 in 6,044

441/440 = 4 cents
176/175 = 10 cents
126/125 = 14 cents

1.05946

7/20 1

15/14 9/8 6/5 5/4 4/3 3/2 8/5 5/3

Figure 3.11. The rational intervals, and the triangles that generate them, can be compared to their equivalent, equal-tempered tones to reveal the metrological tweaks found in historical metrology. These would evidently have had a strong application in their use within harmonic constructions. (There are 1,200 cents in an octave within the system used by modern musicologists.)

*This reminds us of the fact that the excess of the solar year over the lunar, using a megalithic yard of $^{19}/_7$ feet, equals one English foot in the lunation triangle and that the excess of the solar year over eclipse year, as a triangle, will generate the royal cubit as $^{12}/_7$ feet.

We can then see the likely application of geometric metrology, its shortcuts and adjustments, could have made the equal-tempered scale available long before the modern scheme of logarithms came into existence. N : N + 1 triangles not only enabled the capture of essentially logarithmic ratios in celestial time periods, but also provided the logarithmic calculations required by musical mandalas in the round, because the iteration of such triangles inherently lead to logarithmic growth measures. Though such logarithms might not seem accurate, higher accuracies are always available to the builders of larger versions of such triangular calculators, whereas relatively small triangles deliver the accuracy necessary for acoustic comparison to the pure-tone scales.

THE IMPACT ON HISTORY

The numerical encoding of scripture translated the objective speculations arising in the Megalithic period, into a world of literary allusion that became the classical foundations of our civilization. If one had no knowledge of what these scriptures alluded to, then they appeared to have only mythic content, which generated the familiar interpretation of scripture found in the major monotheistic religions: Jewish, Christian, and Islamic.

Octave doubling and tonal experimentation presents the idea that the world is organized by the invariant pattern of the harmonic number field. An octave carries a sense of transformative achievement through the intervals within it that makes new types of plot and characterization possible, outside a musical context and inside a literary one.

The innovation of this new style of storytelling, in the Age of Aries, displaced the numeric realities encoding the structure of the sky. Though the harmonic technicians tried to reconcile the sky to musical harmony, their attempts could find no ideal reconciliation between the cosmic and the harmonic. They found only annoying similarities, themselves remarkable: 365 days were like the 360 degrees of Babylon and the 360 pegs that generate a conceptual calendrical tuning system. (Where 360 is 5 times 72, and 365 is 5 times 73.)

The realization that stories could not only be tonal but also have a circular form might have been the inspiration for the compositional structure evident in the Bible's Pentateuch and later literary works. The fact that ring (or circular) compositional structure has been found to run parallel to the harmonic encoding of documents such as Homer's *Iliad* means that the writers understood both structural and harmonic principles and reverenced both of these literary skill sets. It was probably thought that stories should embody a number of compatible and harmonious steps so that they should return to where they started, though at a higher point (the *octave doubling* of harmonic theory paralleling the *latch* of circular composition).

In the rise of post-classical cultures there was an ignorance of these harmonic and structural literary disciplines and their use in the older scriptures and epics. There was a corresponding drift toward religious dogmatism and a theology based upon scriptural "facts" that supposed such works to be the literal word of God or actual historical events. The growing sense of individualized thinking, which we call rationality, would be suppressed by the convenient dogmatism that God fancied *himself* as a writer rather than being the creator of a universe in which the structure of the cosmos and of musical harmony had led people to write about Him. Ultimately, a rational science would designate all the past systems of thought as merely of historical interest.

The mind machine of human thought ultimately replaced the study of a cosmic machine ordered according to number. This mind machine is based upon subliminal elements from the past—works that survived from antiquity and capabilities that had been won through the previous precessional ages. Since the Middle Ages the growth of the mind machine has been exponential due to new technologies.

However the cosmic and harmonic study projects did not come to an end. Secret groups continued to preserve and evolve secrets about the environmental and inner life of mankind. A harmonic cosmology emerged from a Central Asian group called the *Sarmoun* Brotherhood through the writings of G. I. Gurdjieff before World War I. As we will see in chapters 5 and 6, this emergence unifies the otherwise incompatible views of the cosmic environment and the harmonic theories of chapters 2 and 3.

CHAPTER 4

CRISIS OF THE MIND MACHINE IN PISCES

IN ORDER TO HAVE consciousness, there must be objects to be conscious of, and these must be structured—for only then can the size of our present moment be seen as a living entity that can expand and contract, to remember objects and structures only to forget them again. Consciousness, however, is not a personal possession; it is only the transient achievement of the selfhood that gives rise to it and its activities. In a sense, this transience matches the nature of the existing world in which things come together only to dissipate after a while.

I would suggest that the middle of precession coincides with our time, and we have reached this present moment through the human developments of half of the precessional cycle. In the manner of a storyteller's circular structure or ring composition, we have reached the turn, a point of central meaning. In a story, the present moment is expanded through the actors, objects, and relationships developed, and, by way of these, we can see the central meaning of the entire enterprise. The polarity of the material must then move from the first half to the second: from the collection of meaningful objects to the resolution of their central meaning, and we now have to work more consciously with how meaning shapes our own culture.

In harmonic theory, this opposite point to the start of the cycle is the irrational square root of 2, and this is an irrational tear in the fabric of the existing world through which something—a redemptive power—was thought to enter from another dimension. In the Rig Veda this was the god Agni, the god of fire, for where fire arises, things again move upward. Agni was born of fire, like the fabulous phoenix bird; he, like us, was created not at the beginning of the world, but can arise only at this harmonic point, the turn, within any octave opposite to the note do. This latent force can enter into processes only when the story has been properly developed. Agni was an early prototype for the savior, a redemptive god who is shown nailed upon a world tree or cross, the vertical line of the precessional cycle's symmetry.

To be in the world and not of it: this is a formula in which we are in-between—on the right, desire, and, on the left, non-desire. Whenever this duality is born, Agni comes to burn away the desire for difference and restore the non-desire for wholeness.

Every cycle is a whole, and each whole carries its present moment, which is made of consciousness. Therefore, precession is a single present moment only if it can be grasped, and the fire, kindled at the turn, is a realization of that greater present moment or predicament. Here, at the turn of the precessional cycle, some new fire must be stolen from heaven to establish the next Age. The wisdom of precession is that without precession, there would be no fire, for the precessing pole is a fire stick setting light to successive ages.

The constantly falling pole of Earth is therefore the constantly regenerative nature of Agni; or preeminent god of yoga, Shiva; or the human need for a redemptive work that develops us. The redemption of the present requires the present to become a bigger place for consciousness, and in this we approach some of the fire left by Gurdjieff when he proposed that the God who created the universe had thereby made a constant source of redemption for himself and the creation and had consequently guaranteed the expansion of the existing world. Humans, made in God's image and likeness, are more than an exercise in manufacture; they are a primary cosmic principle on which the universe could be made to achieve new things. To realize this is to realize a vision for the universe that is expanded so there is then no separation of the two sides of the story—of the creation and our role within it.

TECHNOLOGY AND THE
SORCERER'S APPRENTICE

The story of modern science is unintentionally disempowering to the average person, as it attributes life on Earth to being entirely achieved through physical laws, and through the process of natural selection on an accidentally suitable planet. Now that science has developed our technological society, it is technological knowledge and capability that is evolving; we must adapt to it, rather than the knowledge adapting to us. Some expect that through genetic manipulation, we will change our human nature, though we do not know what this will mean; there are so many possible outcomes, and some of these could lead to disaster, which is exactly the scenario painted by Goethe in his poem "The Sorcerer's Apprentice."

> The poem begins as an old sorcerer departs his workshop, leaving his apprentice with chores to perform. Tired of fetching water by pail, the apprentice enchants a broom to do the work for him by using the magic he is not yet fully trained to employ. The floor is soon awash with water, and the apprentice realizes that he cannot stop the broom, because he does not know how.
>
> He then splits the broom in two with an axe, but each of the pieces becomes a new broom and takes up a pail and continues fetching water, now at twice the speed. When all seems lost, the old sorcerer returns, quickly breaks the spell, and thereby saves the day. The poem finishes with the old sorcerer's statement that powerful spirits should be called only by the master himself.[1]

We humans as technologists are now shaped by our own spell— due to various reasons, technologies are arguably beyond the control of the free-market societies in which they develop so rapidly. We can see this in the introduction of a technology as simple as the cell phone, where the original idea, of a mobile telephone, was soon overtaken by its social innovations and new markets, such as texting and photo capability. Meanwhile, cell phones are the cause of car accidents, anarchic mobilizations, economic activities, exposures to radiation, erosion of free time, and so on. It is likely that biological innovations will have unforeseen outcomes, especially when they involve human beings altering themselves.

A common myth in modern society is that of extraterrestrial beings bringing new technologies to Earth, which is then swamped with the

Figure 4.1. "The Sorcerer's Apprentice." Illustration by S. Barth, 1882.

consequences. Technology is such an alien process, but without the aliens: too much of something foreign that comes from outside individuals and changes everything about their lives. All systems need a certain amount of change, but too much change constitutes an imposition from outside, and under such circumstances, systems cannot learn. With more change, they no longer adapt but are changed from outside by a *force majeure*. In this scenario, the active force has moved to the outside, and this eliminates nature as a means of evolution for human beings.

In precessional time, this condition is described as that occurring at the end of any world age: the waters rise up uncontrollably from below, and they threaten the human world, as described in "The Sorcerer's Apprentice." Indeed, the apprentice was not able to learn under the circumstances he had engendered.

The greatest and only intervention from the spiritual worlds is that humans can sense the spiritual and evolve their consciousness through it, because of their innate selfhood. This seems to be a religious idea, yet it doesn't require any religious institution. Because modern science still reacts to its past fight against the dogmas of organized religion, it is especially unreceptive to there being a higher source of order beyond the factual world of material laws. Yet this same science proposes a massive, open-ended revision of human nature and the world through technology and this appears to threaten the evolution of consciousness. We are not being allowed to choose how much technology there will be in the world, because each generation does not really remember what came before and governments go from one crisis to the next.

According to Gurdjieff, the principle on which the universe was founded was that the source of the Creation should not intervene. What intelligent life presents instead is the mechanism whereby higher intelligence can become localized within the Creation. There would be no wisdom if that wisdom was not evolved from experience of the world as actually experienced. But if our experience of the world becomes dislocated, then we will not be able to receive the spiritual world. This risk is very real, as the mechanism for receiving higher intelligence will be incapacitated if humans build a wholly manmade or virtual reality. Nature will then have to develop an alternative intelligent life form should human civilization lose its potential for transforming the spiritual.

It might already be impossible to rescue ourselves from technology as a whole. The challenge is to look for means, such as those technology provides, in the service of our conscious aims. To achieve this, we must restore our sense of purpose as individual meaning makers—in contrast to what centralized mass production and mass consumerism need us to be.

THE MAN OF MANY MEDIA

Compared to animals, it is obvious that human beings are extremely flexible, because we can invent new ways of doing things. Though we use the word media to describe radio, television, Internet, e-mail, and so forth, the real medium is humanity's innovation of new capacities and therefore capabilities. In terms of past chapters: language and stories formed media in which the Neolithic cultural mind could expand to confront issues such as how the world had come to be; metrology and geometry formed media through which cosmic time could be explored; and number notation enabled the harmonic world to be explored as an ideal of mental order.

Media is therefore instrumental in evolving new ideas. It is a genie that belongs to the spaces or intervals between what is already known— as if relatedness belongs to what was called the spirit world. Media is not just those examples that we have created thus far; it is a general principle at work wherever consciousness is involved, for consciousness appears to be a mediating energy that allows sight of something not previously seen.

To mediate between what already exists and what is possible is a type of work distinct from ordinary tasks. Without such a work there are no real possibilities or alternatives to what is otherwise an existential and deterministic world. However, we rely on the exceptional genius to show us new patterns that will then shape us and our culture, and this process implies a deficit—a need for personal creativity to be developed.

In our age, the trigger for our actions is increasingly from the media created by those who came before us. The new media of printing press, books, telephone, radio, television, and internet brings before us a chaos of richly sourced information, which has created a bubble of awareness unprecedented in human history. Whatever the meaning of our age, it must relate to the action of such media. Technology has created new layers of mediation, but how can we be masters and not slaves of media?

The conditions we now experience have been brought about through thousands of years of relentless development in preparation for today's new media. As we have seen in previous chapters, these precursors have added up to a coherent sequence of developments, requiring Ages to achieve them because each step needed a unique focus. It is with such an appreciation of precession that we can perhaps understand what might be our next step, after the current Age of Pisces.

There seems to be a reason for the human being in the middle of all of this, left as we literally are, to our own devices, to see if we can develop something necessary for future growth. The alternative frameworks of the past may come to our rescue, because these frameworks come from ages in which media did not drive human life so relentlessly and humanity still mused upon spiritual worlds. With information about the past, we can look back over the project of consciousness and develop a new sense of purpose regarding how we might use our media.

When attention is refocused on what media *is* rather than its *use,* then it becomes clear that humans are media, because intelligent life is destined to stand between the cosmic and planetary world in a way that a cosmic hierarchy never could. The ancient stories about the sky and precession were the beginning of our taking over the story of

Creation. What we perceive today as our new media, developed through the media of preceding ages, can be extrapolated into a future where humans become a pure medium of intelligible awareness and genuinely free will. Gradually, we are drawing in toward this inner singularity of what we are and employing the subtle media tool of self-knowledge.

THE EMERGING SCIENCE
OF STORYTELLING

Through the action of stories, we humans adopt a characteristic found in all the cosmic bodies that orbit others and rotate about their centers. I propose such rotations and orbits are receptive to cosmic energies; indeed, they are produced by the cosmic dynamism that manifests over Great Time as the precessional cycle, a dynamism with a surprising degree of numerical structure.

In past ages, the alignment to the pole was considered a sacred task—as was the establishment of a sacred center and boundary that represented both the pole and its orbit—by societies seeking alignment to a lawfulness found in the cosmic order. This alignment and establishment of order was achieved through stories, monuments, sacred number sciences, and ritual. A ritual is a repeated story, and the circumstances on Earth of this repetition were invariably organized through places designated as receptive and that often involved temple construction, with geomantic and celestial considerations regarding their location.

It is feasible for modern humans to focus on their own narrative-building skills, not applied to the landscape but by using the ring composition technique introduced in chapter 1. We naturally create stories within lives—a life is a story; biographies tell these. Because the stories from outside are managed—usually full of bad news and centralized restriction—what we need are new foundational stories, as the ancients would have understood their big stories. These, then, create new centers and axes for our world and become worldviews. Modern big stories have lost the sense and meaning of human evolution. The crisis of this age is that there are too few original stories from individuals and too many fabricated ones from large organizations. The ultimate story allows something intelligent to be achieved by the storyteller.

As we have developed reason, we must use reason to receive what is unreasonable—that is, what is creative. The vast world that we have inherited—a heritage made of knowledge—cannot innovate. We have to learn how to make stories not preformed by society and its data banks but coming from the inner structure of the world process—new stories that blend experiences in a creative way.

The task is to capture and process life experiences as if they were coded messages. To transform meaningful experiences into a story is a journey into understanding, a journey first engaged in by a few ancestors when selfhood came of age to the world. What was advanced for a few prehistoric people is now possible for many individuals today.

Almost all the important changes in human society have come through storytelling, but as storytelling floods the world it threatens to turn humanity into a self-enclosed hive of stories to control a collective. Collective storytelling is precipitating a crisis for the mind machine, and we need stories that embody the creative energies that continually arrive on earth—within our life experiences and through synchronicity and relatedness.

G. I. Gurdjieff was very critical of the collective failures of modernity, such as war, and could see that individuals must reconnect with an objective reality where Humans, Nature, and God all require the transformation of individual human beings according to principles built into the universe itself. Born of an Armenian storytelling father, Gurdjieff had roots in the traditional world yet was quite modern in his assessment of the human predicament. He presented the loss of genuine traditional knowledge as the reason for our problems and projected that we could develop a new mode of consciousness—one that would expand upon past traditional forms.

The work Gurdjieff advocated was justified cosmologically by how, at an individual level, humans must transform what occurs to them into a higher meaning—because their experiences were a kind of food that needed to be *digested*. To help us comprehend this possibility, some of Gurdjieff's cosmological ideas are found in the coming chapters in this book. Later in life, Gurdjieff focused on a literary presentation of his powerful ideas and the training of a few individuals, all with the idea of achieving a cultural influence after his death. One of these individuals was J. G. Bennett, who sought to build a full and explicit cosmology based on Gurdjieff's ideas but incorporating ideas from the history, philosophy, and science of his day. Gurdjieff and Bennett indicate how a new precessional imperative could arise as a new tradition.

It is Bennett's line of activity that has taught me, through his student Anthony Blake, the value of story making. The combination of my own numerical work on Megalithic astronomy, Gurdjieff's cosmology, and the ancient art of storytelling formed in me the idea of precession as a developmental cycle. It seemed appropriate to restore a cosmic dimension to story making, because acts of meaning making are a manifestation of the cosmos—they no longer need to be *about* the cosmos. Any practitioner of a new science of story making would understand how Gurdjieff's cosmology usefully transforms our view of the world and the ancient cosmological sciences.

CHAPTER 5

GURDJIEFF'S LAW OF SEVEN

TO UNDERSTAND THE EVOLUTION of consciousness, we must realize that this evolution cannot happen in the sense of it being done to us, even though we appear to experience consciousness. In prerevolutionary Russia, Gurdjieff began his exposition of a unique set of ideas.

> In speaking of evolution it is necessary to understand from the outset that no mechanical evolution is possible. The evolution of man is the evolution of his consciousness. *And Consciousness cannot evolve unconsciously.* The evolution of man is the evolution of his will, and "will" cannot evolve involuntarily. The evolution of man is the power of his doing, and "doing" cannot be the result of things which "happen."[1]

Though every person thinks that they *do,* that they act, almost all of our doing is actually made up of circumstances and influences working through us, and therefore, according to this definition, our consciousness is not evolving. In light of an evolution of consciousness within precessional time, successive cultures largely develop themselves rather than individual human beings, who can evolve themselves only through the power of their own doing. What cultures are providing is a context for individual doing, through their stories of what is important and their developed modes of consciousness.

Gurdjieff gave a structure to how things are done that resembled a musical octave, which he called the law of seven, that not only described human action but was also a cosmological scheme of how God had created the Universe. His law of seven has interesting differences from the concerns of the harmonic specialists of the Age of Aries for he says, "The seven-tone scale is the formula of a cosmic law which was worked out by ancient schools and [only then] applied to music."[2]

This cosmic law of seven is affected by our moon and the ratios it generates, which were discovered in the Megalithic period. The cultural activity of two Ages, involving cosmic study in Taurus and harmonic study in Aries, are therefore brought together in an interesting way within Gurdjieff's own presentation of the law of seven. It may well be that Gurdjieff's ideas are relevant to the ending of our own Age and what needs to be done—in transforming the present world into a new world that can reintegrate how we make meaning and tell new stories—is for us to use some of this Gurdjeffian *fire stolen from heaven,* in the style of a precessional hero.

THE RUSSIAN VERSION

According to P. D. Ouspensky, Gurdjieff first introduces his law of octaves to a group in pre-Revolutionary Russia according to the verbatim report of P. D. Ouspensky in his *In Search of the Miraculous:*

> In order to understand the meaning of this law [of seven] it is necessary to regard the universe as consisting of vibrations. These vibrations proceed in all kinds, aspects, and densities of the matter which constitutes the universe, from the finest to the coarsest; they issue from various sources and proceed in various directions, crossing one another, colliding, strengthening, weakening, arresting one another, and so on.
>
> In this connection according to the usual views accepted in the West, vibrations are continuous. This means that vibrations are

usually regarded as proceeding uninterruptedly, ascending or descending so long as there continues to act the force of the original impulse which caused the vibration and which overcomes the resistance of the medium in which the vibrations proceed. . . . So that one of the fundamental propositions of our physics is the *continuity of vibrations*. . . .

In this instance the view of ancient knowledge is opposed to that of contemporary science because at the base of the understanding of vibrations ancient knowledge places the principle of the *discontinuity of vibrations*.[3]

Something other than music is being referred to and to see this one needs to realize that, with music-making the vibrations are created by an instrument that can achieve many types of note using a purely functional apparatus. Gurdjieff on the other hand is proposing that systems in the universe have vibrations and that the energy of making the notes has to arise through their own inner and outer transformation of their own vibratory level. As with a finger plucking a string, however, these transformations still have to begin with some kind of impulse:

The force of the impulse acts without changing its nature and vibrations develop in a regular way only for a certain time which is determined by the nature of the impulse, the medium, the conditions, and so forth. . . . But at a certain moment a kind of change takes place in it and the vibrations, so to speak, cease to obey it and for a short time they slow down and to a certain extent change their nature or direction; for example, ascending vibrations at a certain moment begin to ascend more slowly, and descending vibrations begin to descend more slowly. In order to determine these moments of retardation, or rather, the checks in the ascent and descent of vibrations, the lines of development of vibrations are divided into periods corresponding to the *doubling* or the *halving* of the number of vibrations in a given space of time.[4]

Because the vibrations are *within actual systems and situations,* and not just carried on the wind, then the language seems obscure in relation to music. Whereas Gurdjieff's theories correspond to musical harmony in theory, the musical intervals involved relate to the processes found in the world itself.

> It has been found and established that in this interval of vibrations, between the given number of vibrations and a number twice as large, there are two places where a *retardation in the increase of vibrations* takes place. One is near the beginning but not at the beginning itself. The other occurs almost at the end.
>
> Approximately:

1000   2000

> The laws which govern the retardation or the deflection of vibrations from their primary direction were known to ancient science. These laws were duly incorporated into a particular formula or diagram which has been preserved up to our times. In this formula the period in which vibrations are doubled was divided into *eight* unequal steps corresponding to the rate of increase in the vibrations. The eighth step repeats the first step with double the number of vibrations. This period of the doubling of the vibrations, or the line of the development of vibrations, between a given number of vibrations and double that number, is called an *octave,* that is to say, *composed of eight.*[5]

The proposal here is that, at some time in the past it was observed that when trying to do things (based on an original impulse), there is a natural point at which the work needs some further impetus. Gurdjieff explains that this further impetus will deflect the original impulse (so as to change the original intent) unless a compatible further impulse from outside is found to revivify the process, in line with the original intent:

The principle of dividing into eight unequal parts the period, in which the vibrations are doubled, is based upon the observation of the non-uniform increase of vibrations in the entire octave, and separate "steps" of the octave show acceleration and retardation at different moments of its development.

In the guise of this formula ideas of the octave have been handed down from teacher to pupil, from one school to another. In very remote times one of these schools found that it was possible to apply this formula to music. In this way was obtained the seven-tone musical scale which was known in the most distant antiquity, then forgotten, and then discovered or "found" again.

The seven-tone scale is the formula of a cosmic law which was worked out by ancient schools and applied to music.[6]

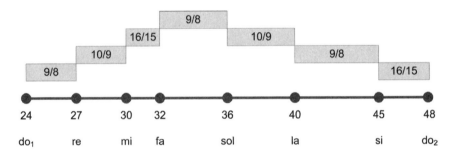

Figure 5.1. The number range 24–48 releases the scale used by cosmic octaves throughout the universe.

This purports to give some historical detail as to the precedence for a form of the law of seven prior to the later work, in the Age of Aries, of a fuller harmonic theory based upon numbers. The irony is that the development of harmonic theory *was itself a deflection* from the original law of seven. Whereas harmonic theory includes all the possibilities of harmony, the Law of Seven explained how things really happened. The law of seven as it applies to the universe was left behind, and its significance forgotten. Gurdjieff further explains:

If we take do as 1 then re will be 9/8, mi 5/4, fa 4/3, sol 3/2, la 5/3, si 15/8, and do 2.

1	9/8	5/4	4/3	3/2	5/3	15/8	2
do	re	mi	fa	sol	la	si	do

The differences in the acceleration or increase in the notes or the difference in tone will be as follows:

between do and re 9/8 : 1 = 9/8

between re and mi 5/4 : 9/8 = 10/9

between mi and fa 4/3 : 5/4 = 16/15 increase retarded

between fa and sol 3/2 : 4/3 = 9/8

between sol and la 5/3 : 3/2 = 10/9

between la and si 15/8 : 5/3 = 9/8

between si and do 2 : 15/8 = 16/15 increase again retarded[7]

Therefore the retardations occur at the semitones of 16/15, when something new is required for the octave to continue.

Because the development of harmonic theory has been important in decoupling, rightly or wrongly, the human mind from cosmic structure, there has been a consequent loss of the original law of seven and its explanation of how our own impulses run out of steam in practice. And we don't even notice this fact even when these impulses clearly achieve the opposite of their original intent.

The same thing happens in all spheres of human activity. In litera-ture, science, art, philosophy, religion, in individual and above all in social and political life, we can observe how the line of the develop-ment of forces deviates from its original direction and goes, after a certain time, in a diametrically opposite direction, *still preserving its former name.*

A study of history from this point of view shows the most aston-ishing facts which mechanical humanity is far from desiring to notice. Perhaps the most interesting examples of such change of direction in the line of the development of forces can be found in

the history of religion, particularly in the history of Christianity if it
is studied dispassionately. Think how many turns the line of development of forces must have taken to come from the Gospel preach-
ing of love to the Inquisition; or to go from the ascetics of the early
centuries studying *esoteric* Christianity to the scholastics who calculated how many angels could be placed on the point of a needle.

The law of octaves explains many phenomena in our lives which
are incomprehensible.

First is the principle of the deviation of forces.

Second is the fact that nothing in the world stays in the same
place, or remains what it was, everything moves, everything is going
somewhere.[8]

While such deviations are to be found within human history, I propose that the precessional cycle is not itself subject any such deviations.
This is because it is a *cosmic process* maintained by the solar system itself.
Therefore, precessional history is on a higher scale in its overall pattern
or form and not subject to the same vagaries that humans experience
on the earth, and precession is probably instrumental in evolving life
on earth.

A HARMONIC COSMOLOGY

The idea that lies behind these vibrations and scales is that the whole
of the universe divides itself according to harmonic principles. That our
own world is affected by octaves with deflections is because the universe
itself comes under harmonic laws by design.

In the big cosmic octave [see figure 5.2], which reaches us in the
form of the *ray of creation,* we can see the first complete example
of the law of octaves. The ray of creation begins with the Absolute.
The Absolute is *the All. The All,* possessing full unity, full will, and
full consciousness, creates worlds within itself, in this way beginning

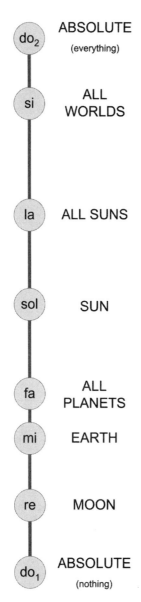

do₂ ABSOLUTE
     (everything)

si   ALL
     WORLDS

la   ALL SUNS

sol SUN

fa   ALL
     PLANETS

mi   EARTH

re   MOON

do₁ ABSOLUTE
     (nothing)

Figure 5.2. The ray of creation organized as an ascending octave scale, according to the above text and adapted from figure 18 of *In Search of the Miraculous*.

the *descending* world octave. The Absolute is the do of this octave. The worlds which the Absolute creates in itself are si. The "interval" between do and si in this case is filled by the *will of the Absolute*. The process of creation is developed further by the force of the original impulse and an "additional shock." Si passes into la which for us is our star world, the *Milky Way*. La passes into sol—our sun, the solar system. Sol passes into fa—the planetary world. And here between the planetary world as a whole and our earth occurs an *"interval."* This means that the planetary radiations carrying various influences to the earth are not able to reach it, or, to speak more correctly, they are not received, the earth reflects them. In order to fill the "interval" at this point of the ray of creation a special apparatus is created for receiving and transmitting the influences coming from the planets. This apparatus is *organic life on earth.* Organic life transmits to the earth all the influences intended for it and makes possible the further development and growth of the earth, mi of the cosmic octave, and then of the moon or re, after which follows another do—*Nothing.* Between *All* and *Nothing* passes the ray of creation.[9]

THE LATERAL OCTAVE OF LIFE

Gurdjieff then went on to explain why life comes about within this ray of creation and how this is reflected harmonically.

And once dwelling on this theme he gave us a diagram of the structure of the octave in which one of the links was "organic life on earth."

"This additional or lateral octave in the ray of creation begins in the sun," he said.

"The sun, sol of the cosmic octave, begins at a certain moment to sound as do, *sol-do.*

"It is necessary to realize that every note of any octave, in the present instance every note of the cosmic octave, may represent do of some other lateral octave issuing from it. Or it would be still more exact to say that any note of any octave may at the same time be any note of any other octave passing through it.

"In the present instance sol begins to sound as do. Descending to the level of the planets this new octave passes into si; descending still lower it produces three notes, la, sol, fa, which create and constitute organic life on earth in the form that we know it; mi of this octave blends with mi of the cosmic octave, that is, with *the earth,* and re with the re of the cosmic octave, that is, with *the moon.*"[10]

The new octave starting with the sun takes its high do as the sun-moon-earth system that manifests as precessional time. The lunar nodal period has a semitone with respect to the Trigon period to form si, and Jupiter's whole tone to the lunar year provides the la-si, with the lunar year forming the medium. Venus provides a descending fifth (3/2) × semitone (16/15) to achieve the rising tone mi (5/4), which is 9/8 × 10/9.

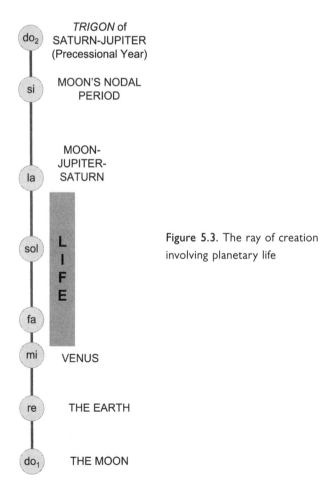

Figure 5.3. The ray of creation involving planetary life

The lateral octave of life, new to Gurdjieff's presentation, is shown shaded (figure 5.4) and is held between the golden mean of Venus, from which life derives its form, and the "King of the Gods" Jupiter, who, along with its conjunctions with Saturn, provides an inner structure for precession of 108 Trigon conjunctions (see chapter 2). Life then has to fill three intervals: the semitone of mi-fa and two whole tones.

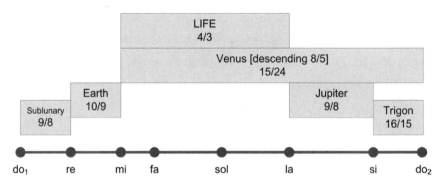

Figure 5.4. The ray of creation for organic life
with regard to its pure tone intervals

This octave is the Earth-centered or geocentric one appropriate to Life in that it is the relative frequencies seen from the earth and not from the sun that generate and maintain this octave. There appear to be credible factual supports for this octave, because the anomalous musical tones found in the geocentric time periods for Venus, Saturn, and Jupiter are available to provide these numerical ratios.

Only precession and the Trigon period relative to it, can provide the necessary time periods over which we can harmonize the lunar year to the eclipse year (within the solar year) and the outer planets, to create one single system, the precessional mill. To achieve this, both Jupiter and Saturn must ring pure tones of 9/8 and 16/15 relative to the lunar year, and Jupiter's cycle relative to the lunar year is then 108 (the traditional number of the moon) times 3/2 lunar months, just as there are 108 Trigon periods within a precessional Age.

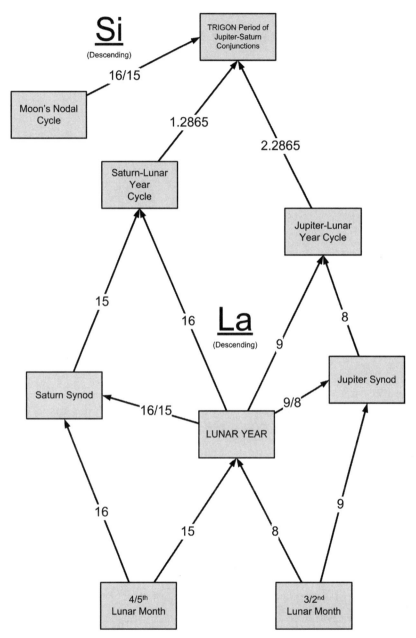

Figure 5.5. The lunar year is synchronous with both Saturn and Jupiter, but Jupiter holds the whole tone, because it has deposed Saturn, leaving it with a halftone. Both tones have a cyclic period, and each period exhausts itself to divide into the Trigon period of Jupiter-Saturn conjunctions, from which come the note la. The note si is provided by the moon's nodal cycle, just 15/16 of the Trigon period.

Venus exceeds the solar year by a Fibonacci ratio of 8/5, which corresponds to a descending pure tone of 24/15 to reach the note mi.

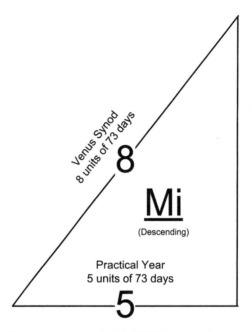

Figure 5.6. The Venus synodic period is 583.92 days relative to the practical solar year of 365 days, which makes the triangle 8/5 (within better than 1 part in 7,300). (The practical year being the whole numbered year, in days, that relates to living on Earth.)

The first interval for life is mi-fa, and we know that the molecules of life are found in galactic molecular clouds and were in the oceans of early Earth. The key ingredient is the replicating molecule DNA. Each cycle of DNA has a Fibonacci relationship between its length and width, such as the 8/5 found in the Venus synod or the 13/8 found in the Venus orbit relative to the orbit of Earth.[11] Venus supports the necessary interval containing the golden mean, in Fibonacci form from which Fibonacci ratios are derived using any two adjacent integers. The golden mean proportion is also found throughout life and distinguishes the living from the nonliving world.

We can imagine that vegetation and animal life are involved in the succeeding two intervals and that the giant planets lie beyond these, in the realm of the gods but related to the human world. Within John Michell's diagram (adapted)[12] of the sublunary world (figure 5.7), the center of the moon is a whole tone greater than the polar radius of Earth to 1 part in 99, in which the center of the Moon describes a circle with the same perimeter as the square inscribed by the mean Earth.

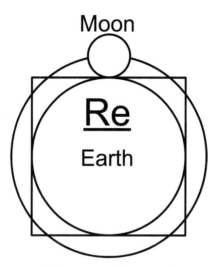

Figure 5.7. The expansion of the moon relative to the mean Earth approximates to the whole tone of 9/8 and has a perimeter of the squared Earth, the moon expressing in its circumference the deficiency of the inscribed circle from the square, according to Pi as 22/7.

So much for the cosmic octave that surrounds life, but life itself is then subject to the moon's alteration of the law of seven.

A REVISED LAW OF SEVEN

One of Gurdjieff's reasons for explaining octaves to his students was that they might come to see, and learn how to overcome, the retardation or deflection caused by semitones. This has to be done alone, remembering from above that "consciousness cannot evolve unconsciously." We do not attempt here an exposition on Gurdjieff's practical teaching methods, since the aim is to relate Gurdjieff's teaching to the earlier ages and, where possible, provide some new and interesting understandings relative to the law of seven.

Gurdjieff wrote more about the law of seven decades later, perhaps having refined how this should be presented, within his cosmological epic *Beelzebub's Tales to His Grandson*. At this later date, he called the law of seven the sacred Heptaparaparshinokh. Its definition begins this way:

> Our COMMON FATHER OMNI-BEING ENDLESSNESS, having decided to change the principle of the maintenance of the existence of this then still unique cosmic concentration and sole place of HIS most glorious Being, first of all altered the process itself of the functioning of these two primordial fundamental sacred laws, and HE actualized the greater change in the law of the sacred Heptaparaparshinokh.
>
> These changes in the functioning of the sacred Heptaparaparshinokh consisted in this, that in three of its *Stopinders* HE altered the, what are called "subjective actions" which had been until then in the *Stopinders,* in this respect, that in one HE lengthened the law conformable successiveness; shortened it in another; and in a third disharmonized it.
>
> And, namely, with the purpose of providing the 'requisite inherency' for receiving, for its functioning, the automatic affluence of all forces which were near, HE lengthened the *Stopinder* between its third and fourth deflections.
>
> This same *Stopinder* of the sacred Heptaparaparshinokh is just that one, which is still called the "mechano-coinciding-Mdnel-In."
>
> And the *Stopinder* which HE shortened, is between its last deflection and the beginning of a new cycle of its completing process; by this same shortening, for the purpose of facilitating the commencement of a new cycle of its completing process, HE predetermined the functioning of the given *Stopinder* to be dependent only upon the affluence of forces, obtained from outside through that *Stopinder*

from the results of the action of that cosmic concentration itself in which the completing process of this primordial fundamental sacred law flows.

And this *Stopinder* of the sacred Heptaparaparshinokh is just that one, which is still called the "intentionally-actualized-Mdnel-In."[13]

In a nutshell, the two semitones were altered, and a third effect was that between the first semitone and the second at the end of the octave, a third interval has become disharmonized, because it lies between these changes. "As regards the third *Stopinder,* then changed in its 'subjective action' and which is fifth in the general successiveness and is called 'Harnel-Aoot,' its disharmony flowed by itself from the change of the two aforementioned *Stopinders*."[14]

Gurdjieff moved from a technical description toward a strange combination of exact articulation and challenging sentence structure with the aim of requiring the reader to exert an extra effort of concentration (a "subjective action"). He also moved to the more involving and flexible story form to make an epic tale, though often dismissed as preposterous fantasy.

His new view of the octave as seven *intervals,* rather than eight notes, was significant, because his law of seven was about the work of transformation, rather than about music. It is significant that he calls these seven intervals Stopinders in his sacred Heptaparaparshinokh.

Gurdjieff's description of the law of seven had itself evolved. Indeed he says, ". . . for almost forty years, day and night, I have persistently studied these great laws of world vibrations until the understanding of their meaning and actualization has become for me, as it were, second nature."[15]

Gurdjieff appears to lengthen and shorten the diatonic semitones of the major scale—the 16/15 ratios that were presented in his Russian version as the universal law of seven. Though Beelzebub says that God did this for the whole universe, clues indicate that on planets with a large moon, special things happen due to such moons—including the arising of life and intelligent life. Gurdjieff now presented the law of seven as it presents itself to us on Earth, rather than as the universal cosmological version he presented in the Russian period.

THE MOON CHANGED THE LAW

According to Beelzebub, our moon was created by an accidental collision with a comet called Kondoor, an explanation that fits with current scientific theories outlined at the end of chapter 2. (See Intermezzo: A Moon that Created the Earth, page 64.) Because of this "accident," the cosmic harmony was threatened and a panel of cosmic specialists were assembled to decide how to harmonize the resulting moon.

> And they resolved that the best measure in the given case would be that the fundamental piece, namely, the planet Earth, should constantly send to its detached fragments, for their maintenance, the sacred vibrations "askokin."
>
> This sacred substance can be formed on planets only when both fundamental cosmic laws operating in them, the sacred "Heptaparaparshinokh," and the sacred "Triamazikamno," function, as this is called, "Ilnosoparno," that is to say, when the said sacred cosmic laws in the given cosmic concentration are deflected independently and also manifest on its surface independently—of course independently only within certain limits. . . .
>
> About how and why upon planets, during the transition of the fundamental sacred laws into "Ilnosoparnian," there arise "Similitudes-of-the-Whole" and about what factors contribute to the formation of one or another of these, as they are called, "systems of being-brains," and also about all the laws of World-creation and World-maintenance in general, I will explain to you specially some other time.[16]

Complex life, or similitudes-of-the-whole, arise only on planets that have a large moon, because the law of seven is chaotically deflected by the variety of impulses and conditions that occur within life, as beings interact. This interaction affects the mechano-coinciding-Mdnel-In, which is the semitone between mi and fa, obviously lengthened from the cosmic semitone of 16/15 by the presence of the moon itself.

> Now, my boy, in my opinion, before going on to a more detailed explanation of how their higher-parts were then coated and per-fected in the common presences of the first Tetartocosmoses, as well as in the common presences of those who were afterwards named "beings," it is necessary to give you more information about the fact that we, beings arisen on the planet Karatas, and also the beings arisen on your planet called Earth, are already no longer such "Polormedekhtic" beings as were the first beings who were transformed directly from the Tetartocosmoses, . . . but are beings called, i.e., nearly half-beings, owing to which the completing pro-cess of the sacred Heptaparaparshinokh does not proceed at the present time through us or through your favorites, the three-brained beings of the planet Earth exactly as it proceeded in them. And we are such Keschapmartnian beings because the last fundamental *Stopinder* of the sacred Heptaparaparshinokh, which at the pres-ent time almost all the beings of the Megalocosmos call the sacred "Ashagiprotoëhary," is not in the centers of those planets upon which we arise—as it occurs in general in the majority of the planets of our great Megalocosmos—but is in the centers of their satellites, which for our planet Karatas is the little planet of our solar system which we call "Prnokhpaioch," and for the planet Earth, its former frag-ments now called the Moon and Anulios.[17]

The importance of this point is highlighted by its convoluted delivery of the facts: The first Polormedekhtic beings in the megalo-cosmos were different to the Keschapmartnians, such as ourselves, who live on a planet in which the last fundamental Stopinder of the sacred Heptaparaparshinokh occurs in the centers of their satellites rather than in the centers of the planets in which those beings arise. The idea of a lateral octave of life has been reset into a far richer picture in which the moon's harmonization has been crucial, because this affects how life develops—life will be similar to that already arisen (that is, to the cosmos and its harmonies). Disharmonies will have consequences down the line that could affect the cosmos. This can be illustrated in the danger of wrongly developed Keschapmartnian beings who inflict great evils upon the population but are then consigned to specially disharmonized planets that have moons, and on these planets nothing can be achieved.

THE PROBLEM OF OBJECTIVE EVIL

Beelzebub calls those intelligent life-forms that develop disharmoniously Hasnamuss. As "similarities to the whole," they are perhaps relics of the disharmonious past of the moon. They are carried forward in the sense that in the past, human beings were arrested in their development by the inclusion of an organ called *kundabuffer,* which was *designed* to make early humans lack a conscience—that is, they served the moon in its development rather than their own evolution. In the past, humans were adjusted in this way to provide only the energies required to establish the moon in its present harmonious condition. This means that there are higher energies that are naturally produced by life and that actually contribute to the local cosmic harmony.

When the moon achieved harmony, the organ kundabuffer was removed by the same cadre of archangels who had introduced it, but the effects lived on through the habitual consequences of the organ, that were inherited. This is Gurdjieff's original sin, which originated with higher beings rather than the human race, all due to these *unforeseen consequences* of the organ kundabuffer when removed.

Some such Hasnamuss individuals develop themselves spiritually, and they can thereby inflict harm and suffering upon others through their developed powers, according to the law of seven. Such wrongly developed people are described as passing on to special planets after death—places where eventual redemption is possible but also one place from which there can be no escape: "One of these four disharmonized planets called 'Eternal-Retribution' is specially prepared for the 'Eternal-Hasnamuss-Individuals' and the other three for those 'Higher being-bodies' of Hasnamusses in whose common presences there is still the possibility of 'at some time or other' eliminating from themselves the mentioned maleficent something."[18]

In a sense, then, the Hasnamuss goes to the place corresponding to his or her own disharmony with the cosmos, a planet that has a disharmonized moon exactly like the condition of Earth itself when the maleficent organ kundabuffer was introduced into humans so that they would harmonize the moon.

The overall message is that harmony rules the cosmos—but as a principle hidden in everyday life, except through the values practiced by human beings. This is not far from the notions of harmony found in the works of the Age of Aries, that harmonious intervals are good and disharmonious intervals are evil. The problem for contemporary society is the invisibility of the law of seven as a cosmic rather than as a musical value system. Gains in modern musical appreciation cannot make up for humanity's loss of the law of seven as a living principle, and Gurdjieff sought to re-establish this, as a tradition lost in precessional time.

A TONE-MANDALA FOR HEPTAPARAPARSHINOKH

It is difficult to form a picture of the Heptaparaparshinokh version of the law of seven, and perhaps Gurdjieff intended this with his labyrinthine prose. By describing intervals rather than notes, the law is structurally reciprocal to the usual do-re-mi octave. As we have seen, his move to using intervals was necessary because the intervals are processes of transformation rather than ratios between musical notes. If we apply the harmonic mandalas of McClain, the form of Gurdjieff's law becomes clearer. We can start with the cosmic diatonic major scale that Gurdjieff proposed as the universal scheme during his early talks.

There are seven intervals and "HE lengthened the *Stopinder* between its third and fourth deflections"[19] and "the *Stopinder* which HE shortened, is between its last deflection and the beginning of a new cycle of its completing process."[20]

The ratio of the lunar month to the lunar orbital period is 27/25. This is a lengthening of a semitone of 16/15 by 81/80, called the syntonic comma, which relates the pentatonic and just tunings. This forms the new mechano-coinciding-Mdnel-In with the purpose of "providing

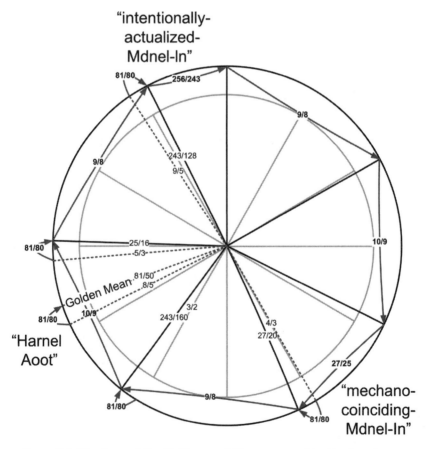

Figure 5.8. The form of Gurdjieff's sacred Heptaparaparshinokh after the moon has adjusted the two semitones within the diatonic scale

the requisite inherency for receiving, for its functioning, the automatic affluence of all forces which were near"[21] based on the moon.

If we recall that an octave must exactly double, then reducing the final semitone of 16/15 by 81/80 yields a new final Stopinder of 256/243, the Pythagorean *leimma,* which is the ratio of the solar year to the eclipse year. This follows the form of the "'intentionally-actualized-Mdnel-In' . . . for the purpose of facilitating the commencement of a new cycle of its completing process."[22]

In the third changed Stopinder, called Harnel Aoot, "disharmony flowed by itself from the change of the two aforementioned *Stopinders,*"[23] as can be seen from the fact that between the third and last Stopinders the two notes we call la and sol must, as notes, also be shifted by 81/80. The relationship to the early development of the scale has been changed rendering this, in the area of achievement, more difficult harmonically.

The known ratios of the moon and Beelzebub's statements about planets with large moons appear to offer a new form for Heptaparaparshinokh that follows many of the rules of harmony but generates an unusual scale.

INTERMEZZO

GURDJIEFF AS AVATAR

GURDJIEFF WAS NOT WHAT he has come to represent within our culture. His book *Beelzebub's Tales to His Grandson* is the autobiographical account of a being who descended to Earth on a number of occasions in order to see what was going on and possibly, to intervene. This is the behavior of an *avatar*—an Indian term, largely for the god Vishnu (a galactic god) who comes to Earth to rid it of false demonic cultures and to support the dharma of cosmic order. As a turtle, for instance, one of these incarnations of Vishnu upholds the churn that the gods and demons turn, and the churning of the oceans is one of the clearest symbols of the polar mill that we have in myth. Was Gurdjieff a precessional hero?

John G. Bennett says, "Gurdjieff spoke ambiguously about himself. Sometimes he came very near to claiming that he was an *avatar,* a Cosmic individual incarnated to help mankind."[1] We must keep in mind that presenting himself as an avatar could engender serious problems in its effect on other people: the claim has proved destructive for many, irrespective of their true status. In addition, in the case of an avatar the experience of being incarnated is not without its confusions and ambiguities. In a sense, the term *avatar,* meaning "a descent," must be used in a fresh way.

Gurdjieff's primary question was: *What is the sense and significance of life on Earth in general and of human life in particular?* Bennett suggests that

> The answer Gurdjieff gives to the question is radically different from any current views. Gurdjieff asserts in *Beelzebub's Tales* that the doctrine of reciprocal maintenance is derived from "an ancient Sumerian manuscript" discovered by the great Kurdish philosopher Atarnakh. The passage quoted runs: "In all probability, there exists in the world some law of the reciprocal maintenance of everything existing. Obviously our lives serve also for maintaining something great or small in the world."
>
> This passage occurs in the description of a Central Asian fraternity called "The Assembly of the Enlightened," which had existed from Sumerian times and flourished openly in the Bactrian kingdom when Zoroaster was teaching. After Zoroaster, it disappeared for a hundred generations and only now begun to send out into the world its "Unknown Teaching." I have suggested that this is the Sarman society [one of Gurdjieff's direct sources].[2]

Beelzebub goes on to say that "the suppositions of the Kurd Atarnakh were very similar to the great fundamental cosmic law *Trogoautoegocrat*."[3] This law prevented the creator from being subject to the merciless Heropass (or time), once the reciprocal feeding of everything that exists (in time) ensured the permanent harmony of the universe.

Gurdjieff's vast harmonic cosmology emerges from a simple principle that, once implemented, would set up a universe in which octaves can "beg, borrow and steal" from each other as a necessary way to progress. Surely, such a simple principle in everything around us should be obvious.

Bennett makes the observation, which is also true for modern science,

that very little attention is given by Eastern religions and philosophy to *purposes* and so do not need to account for anything.

Buddhism, in all its forms, rejects such questions as futile and that the aim of existence is man's own need to escape from *durkha*, which does not mean suffering so much as the conditioned state of the incarnated self. The one significant exception is the old religion of Zarathustra, which taught that both life on the earth and man endowed with intelligence were created to be allies for the Good Spirit Ahura Mazda in the struggle with the power of darkness. The Avestan hymns are full of references to the role of man as a helper in the cosmic process. For example Yastna 30.9 has the invocation: "May we belong to those who renew the world and make it progress!" I have given my reasons for believing that Gurdjieff found the Zoroastrian tradition lived on in Central Asia long after it ceased to be a state religion. For some reason, this important myth was forgotten and for a very long time the question "why does life exist on the earth?" was lost to view.[4]

It may well be that the highest teachings always need some kind of a renewal on Earth by an avatar, because such teachings are simpler than all others. The simplicity of the trogoautoegocratic principle is that a similar form is employed throughout the universe (the laws of seven and three) in order to achieve the same principle throughout (the principle of reciprocal maintenance). The rest of what happens is simply consequent and subsequent.

CHAPTER 6

MULTICULTURALISM AND THE END OF TRADITIONAL KNOWLEDGE

WHEN GURDJIEFF ARRIVED IN Russia before World War I, he had traveled extensively to the East, where he collected and developed practices and ideas both old and new. He could see the effects of modernity—influences were becoming ever more blended and traditional boundaries blurred. His vision of vibrations as an ancient tradition of the law of seven was a perfect vehicle for a synthesis of traditional knowledge and the modern idea of lawful processes.

The primary components of Gurdjieff's vision are the laws of blending and of process: the laws of three and seven that form a single cosmological unity. This cosmic view, and his contemporary critique of "the terror of the situation" due to the organ kundabuffer, gave perspective on what the world needed—the harmonious development of humans. Gurdjieff spent twenty years teaching people, using challenging techniques, not only to abandon institutional teaching altogether but also to continue to look for a means to achieve his aim, of restoring what modernity had taken from human beings: "Today, civilisation, with its unlimited scope in extending its influence, has wrenched man from the normal conditions in which he should be living."[1]

129

GURDJIEFF, THE CULTURAL DOCTOR

The focus of Gurdjieff's work with students was psychic integration based on their efforts and his techniques. Modernity, he believed, is disorganizing to the psyche, which is asleep to its true potential (that intelligent life has the power to recreate itself in the form of its meaning). He also thought that what occurs in cultures is what is found in the individual but on a larger scale.

Gurdjieff drew attention in a new way to what humans eat: the first food is already familiar, comprised of what we eat and drink. The second, less familiar food has a cosmic origin; these are found in the air but are rarely digested. The third food is the continuous stream of impressions that can similarly contain higher energies within them. In the course of a human life, these three foods can interact according to the law of seven, producing higher substances that can form a "higher being body" in a type of alchemical process.

Modern life increasingly interferes with psychic integration—the transformation of higher substances—and therefore with the formation of higher being bodies. The reason for this individual crisis is that, within modernity at this transition between ages, we have largely lost any traditional knowledge about this transformation and need a modern equivalent.

Chapter 1 proposed that the first half of a story manifests what is to be known to the selfhood in the second half. This is similar to a person coming to a realization after having had time to "digest" information as a type of food from which more can be made. Today we have a great deal of general knowledge, and all this knowledge is part of our story but only some of it is necessary in forming the path of return—the path through which self-knowledge emerges from knowledge in general. Our information bubble has been created, through humans but by nature, so that the universe *might* become self-aware through intelligent life.

The first of Gurdjieff's higher being bodies (which he called the *kesdjan* body), is formed from the integration of our experiences—that

Figure 6.1. Alexander Salzmann's design for the program of
Gurdjieff's Institute of the Harmonious Development of Man

is, from our impressions—hence these impressions are described as a
third food. Just as the arrival of larger groups and communal values created new outer languages, the kesdjan body requires the formation of
an inner language in which we can digest the creative meaning of our
environment and develop consciousness.

A human life collects experiences that are either stored or further
assimilated. Stored experiences are incoherent, and become food for the
moon, the creation. On the other hand, by understanding and assimila-
tion these experiences can become food for the soul.

The crisis of multiculturalism is that it cannot provide for our traditional needs or innovate a new tradition for multinational consumerism: the forces of incoherence and distraction assert their gold standard of mass consciousness. Gurdjieff—decoupled from the preconceptions associated with spiritual teachers—can be seen as a cultural doctor and exemplar for a turning point: diagnosing the modern condition (from a traditional perspective) and prescribing a contemporary course of treatment.

THE SUBJECTIVE EFFORT OF LEARNING

A universal law of seven gave rise to the universe, and the subsequent alteration of this law by the creation of the moon was harmonized—its semitone intervals becoming lengthened and shortened.

> These changes in the functioning of the sacred Heptaparaparshinokh consisted in this, that in three of its *Stopinders* HE altered the, what are called "subjective actions" which had been until then in the *Stopinders,* in this respect, that in one HE lengthened the law conformable successiveness; shortened it in another; and in a third disharmonized it.[2]

The "subjective actions" of this change can be equated to the addition of 81/80, the syntonic comma, to the third Stopinder. There must then be an identical reduction by the same amount in the seventh Stopinder and a consequent dislocation of the notes between these two changes, affecting the fifth Stopinder. Because these changes are subjective, they involve how, *from our point of view,* these Stopinders are altered in their working.

Because we are talking an environment created by the moon, the extra subjective action required relates to (a) the extremely rich environmental conditions and (b) our intelligence in seeing what these mean—more than would otherwise be required. This new law of seven is about processes on Earth, and any who carry out processes must then conform to the requirements of the changed law. Extra actions are required within the law's new third Stopinder, called subjective actions because we must provide them.

THE ORIGINAL IMPULSE

A suitable start for the law of seven is the lunar day—that is, the time taken for the same part of Earth's surface to view the transit of the moon in the sky. This is the day responsible for two tides. The lunar day is 24.846389 hours long, about 50 minutes longer than a solar day. The sidereal day is 23.934444 hours long, and the lunar length of day ÷ the sidereal length of day results in the fraction 1.038^{101754}. This ratio is 1 part in 125 less than 25/24, an important sub-ratio in harmonic octaves. Gurdjieff says: "The influence of the moon upon everything living manifests itself in all that happens on the earth. The moon is the chief, the nearest, and immediate motive force of all that takes place in organic life on earth. All movements, actions, and manifestations of people, animals and plants depend upon the moon and are controlled by the moon."[3] (See figure 6.2.)

The first tone in the evolutionary law of seven is 9/8, which is made up of 25/24 × 27/25. This means that the starting note can be constructed from the ratio of lunar to sidereal day but expanded by the cubing of the altar ratio 126/125 and then by the ratio of the lunar month to the lunar orbit: 27/25. This means the moon could be the cause of all new impulses on Earth through the ratios found between its key time periods—manifesting as synchronicity in the apparently chaotic interactions within environments.

In the second Stopinder, the impulse enters a preparatory stage seen in how we respond to our impressions by organizing them. The universal law remains unchanged in these two Stopinders relative to the cosmic octave.

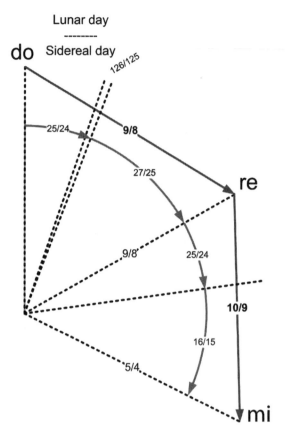

Figure 6.2. The first two Stopinders in the octave of doing. The intervals can be seen as having semitone possibilities, which, in chapter 7, we will see as representing two different aspects in the achievement of each whole tone. (The presence of semitone intervals within whole tone intervals is touched on in *In Search of the Miraculous*, page 126, and might yield extra insights.)

MAKING SOMETHING OUR OWN

The third Stopinder was lengthened by the presence of the moon, and in recent geological time the lunar month–to–orbital ratio came to be 27/25. Over this period of the development of the lunar orbit, life on Earth has evolved because of the moon and its replacement of the planetary interior as terminus for the ray of creation or cosmic octave.

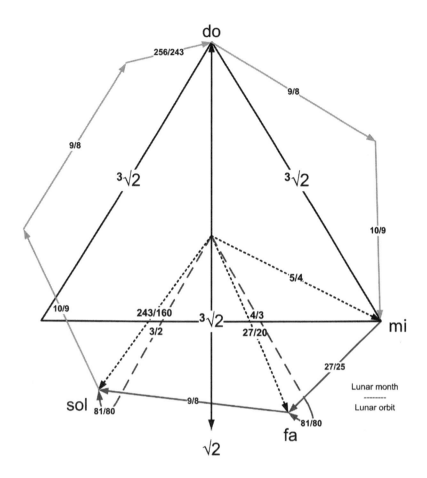

Figure 6.3. The third Stopinder of the sacred Heptaparaparshinokh, which was altered by the moon, and the fourth Stopinder. Together, these bridge the two important roots of 2, the cube root ($^3\sqrt{2}$) and square root ($\sqrt{2}$).

This third Stopinder occurs just as the first third of octave doubling is nearly complete at the note mi of 5/4. We know that an additional 126/125 turns 5/4 into a solution to the ancient problem of "doubling the volume of the cubic altar of Apollo." Schwaller de Lubicz suggests that, "the Ancients begin with volume, from which to proceed their own definitions of surfaces (planes), edges (lines), and vertices (points) . . . [the Egyptians having] a system of 'volume development' established on the principle of proportional growth."[4]

The octave note at the start of this third Stopinder is mi, which is 5/4. This Stopinder crosses the cube root of 2, which has the esoteric connotation of the doubling of the cubic altar's volume. The idea arises of *three different dimensions* being present, in a tone-mandala, as the three points of an upward facing equilateral triangle (figure 6.3)—as found in Gurdjieff's enneagram diagram. This doubling in volume runs parallel to the octave doubling in harmonic theory.

Innovative physicist Peter Rowlands says, "The doubling effect is natural due to the basic concept of duality, and three-dimensionality is itself related to the real/imaginary distinction and the process of multi-dimensionality by doubling is very subtle."[5]

In this way, invisible dimensions are involved in world laws but are found *collapsed* into the one dimension of music. This means that the work of achieving octave doubling involves the inner and outer life and the opening of three spatial dimensions, within which work really takes place on Earth. We are easily lost in the subjectivity these dimensions create when collapsed onto the flatland of the everyday. Dimensional objectivity appears to be a learned skill.

Vibrational properties must build on one another to achieve the result of octave doubling and become stable in their own right. This process is different from that of musical notes, which exist in their own right—and this is the difference between the musical notes of *In Search of the Miraculous* and the later presentation of intervals in *Beelzebub's Tales to His Grandson*.

At the heart of the fourth Stopinder (an interval of 9/8) lies the middle point of the octave, which, as a note, is the square root of 2—the geometric mean of octave doubling. In a ring composition, this is considered the central meaning of the story, and in the Vedic tradition it is associated with the created god Agni. The square root of 2 represents the interrelation of two dimensions in space (as we see in the 1-1-$\sqrt{2}$ triangle) and is, therefore, the first irrational, and only approachable by larger integer fractions. There are known traditional approximations: using seven, 10/7 is 1.4<u>28571</u> rather than the actual 1.414214; and using seven and eleven, 140/99 is 1.4̲1̲, which is accurate to 1 part in 19,600 (meaning it is very accurate). This latter approximation came to be referred to as "giving back the hundredth part." The former approximation is produced by Plato's city known as Magnesia, in which 5,040 is 720 × 7, meaning that the 360:720 octave can generate the 10/7 approximation.

According to Bennett, this is where contact with the medium you are working with takes place. It is where the worker interacts with the problems and opportunities presented by the real world. Something is being done that will make *the turn* toward the evolutionary left side of the octave (using the language of ring composition).

The new perspective: both the square root of 2 and the cube root of 2 are vital. It becomes clear, then, that in order to work, something must enter from below—must come from the cracks within the created world and not from the world of rationality. The rational gods and the irrational demons form a true dynamism for the universe in which our word *rational* rightly points to both our rationality and the integer ratios of harmonic theory. These two worlds, mental and sensory, form a necessary dichotomy for the octave itself. These gods and demons are not the God who created the universe, but are the created *world of possibility* that the gods and demons animate.

GOING BEYOND OURSELVES

The fifth Stopinder, the Harnel Aoot, is disharmonized by the lengthening of the third Stopinder, and the effect of this is that the natural note 8/5 is no longer central, and this affects the fifth Stopinder's relationship to the golden mean (see figure 6.4).

The Venus synod of 8/5 (1.60), which occupies the middle semitone, must shift by 81/80 in the new arrangement—and this is 81/50 or 1.62, which is very close to the actual golden mean of 1.6180. The moon has achieved this golden mean: there are exactly 20 lunations in a golden mean year of 1.618 years (see figure 6.5).

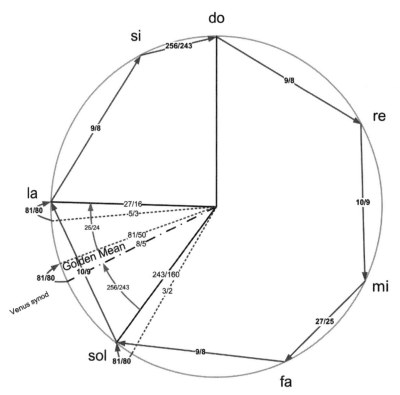

Figure 6.4. The octave is affected most disharmoniously in the area of the fifth Stopinder, where a rational golden mean of 8/5 would sit. To compensate, the Venus synod of 8/5 is extended by the moon to achieve 1.62, 81/80 greater.

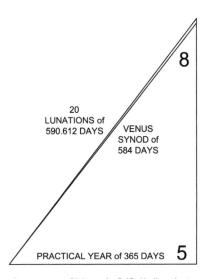

Figure 6.5. The adjustment of Venus's 8/5 (1.6) relation to 81/80 greater yields the golden mean. Adapted from figure 3.5 of my book *The Matrix of Creation*.

The common unit of change in the Law is 81/80, which is available through the common change to the lengthened third Stopinder and the shortened final Stopinder. It is achieved by the 27/16 note, which completes this fifth Stopinder. In this way, there are three inner intervals—the first arises from the same interval as the final Stopinder (256/243), the second (25/24) is provided by the cosmic number field in the original law of seven, and the third arises from the differential lengths of the semitone Stopinders, namely the syntonic comma of 81/80.

Bennett describes the challenge of the fifth Stopinder.

> [Gurdjieff] speaks in an extraordinary way about [the fifth stage], saying that it can give "results not equal but opposite to each other" and "It corresponds to the moment of truth. There is an easy way out, the path of compromise: . . . we can either enjoy it or serve it. At this point we can make things easy for ourselves and it is very tempting to do so. We have reached the point at which we have already created something and only we ourselves can decide what is to come of it. It is possible in this moment to give or to take. We can be captured by what we are doing and at the same time imagine that we are gaining and that we are the master in control. The reality is just the opposite: we are slaves. Only when we leave ourselves out of it and serve the process can we become masters. We become true masters by making ourselves slaves but when we try to become masters we become slaves. If we use our creative power for ourselves it destroys us. If we sacrifice ourselves in order to serve the creative power, then it creates us."[6]

In the next chapter, we see that spiritual values enter into this fifth Stopinder, but also a change in the "subjective action" has occurred with respect to how to perfect something.

The sixth Stopinder is associated with transformative substances. Within the body, the substances of sexual reproduction allow us to go beyond ourselves as bodies in forming newly arising bodies. On another level, and in tune with our personal achievements to fulfill these Stopinders, there arises *helkdonis,* which is the substance from which humans can generate their higher being bodies or souls. The helkdonis gives the appearance of alchemy—some kind of gold is produced that enables spiritual development. The idea is that human work, approached from the point of view of serving the actual conditions of the physical world—the earthly and cultural environment—can generate spiritual energies that are not associated with normal ideas of God, gods, or demons, and this can become a higher intelligence on the planet. This work could not be done to or for human beings by the cosmos but must be done by humans within creation, namely on planets with large moons.

The seventh Stopinder has the same interval as that between the eclipse and solar year. This ratio, within the year, is exactly that as

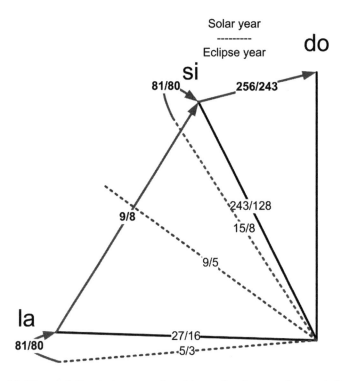

Figure 6.6. The sixth Stopinder naturally achieves the shortened seventh Stopinder, provided by the eclipse–to–solar year ratio.

between the Saros and Metonic cycles (19 eclipse years and 19 solar years, respectively). The interval of 256/243 is reduced from 16/15 by the subjective action of 81/80 so that what one would normally have to do is eliminated altogether from the interval, because,

> [Our COMMON FATHER OMNI-BEING ENDLESSNESS] pre-determined the functioning of the given *Stopinder* to be dependent only upon the affluence of forces, obtained from outside through that *Stopinder* from the results of the action of that cosmic concentration itself in which the completing process of this primordial fundamental sacred law flows.[7]

This means that this intentionally-actualized-Mdnel-In receives this name because it is the automatic completion of the intention of the Stopinders that precede it. As Bennett puts it: "What we have to do to cross the gap is put ourselves in the position of the result. The result is called by [Gurdjieff] *Resultzarion*."[8]

This interval of completion on Earth appears as the Saros and Metonic cycles, which hold a very perfect pattern of completion within 19 years. The years are themselves made up of lunar months and orbits, and they are exceeded by the Venus synod, which is further extended by a golden mean year. The moon is evident everywhere in this local cosmic framework, including how new beginnings come about.

MEGALITHIC ORIGINS OF GURDJIEFF'S LAWS

In this way, Gurdjieff's ideas appear to correspond to the cosmic constants of the moon, which, if Gurdjieff's harmonic cosmology is correct, means that life on Earth poses definite challenges for individual human beings—something our present civilization cannot see or address.

Gurdjieff claimed that in the past this law was elucidated and articulated in symbolic forms he calls Legominisms. Only some of this knowledge survives today—within monuments, symbols, and stories—

but to understand these requires consciousness. Alongside the loss of knowledge, the integrated powers of the human being have degraded so that most people live different lives of the head, heart, and body from those lived long ago. New forms of knowing and being have displaced the old knowledge and ways of being. Most of these changes started in the Age of Aries, when an original law of seven was displaced by the study of musical harmony: a diatonic scale altered by the moon would seem irrelevant to the students of musical harmony.

If a harmonic explanation of cosmic structure had developed, then an adjusted diatonic scale might have been developed using the most obvious diatonic scale as a starting point. There then would have been diatonic music using metrology to define string lengths.

In Greek myth, Apollo is credited with inventing the seven-stringed lyre. Apollo was said to come to Britain from Greece on the occasion of his nineteenth year—and 19 is the number of years in the Metonic cycle and the number through which Megalithic astronomers derived the astronomical megalithic yard of 19/7 feet.

As detailed early in chapter 3, it is likely that octaves were first built using the triangles in the Megalithic to form their string lengths. Harmonic tones can easily be constructed using triangles, and metrology probably turned to such use—for many historic measures are harmonic intervals relative to the English foot.

A further reason for this approach is that within the developing consciousness of the cosmic ratios, the solar year–to–eclipse ratio is accurately the Pythagorean leimma of 256/243, which is evident in the triangle for the eclipse–to–solar year: it has sides that are 18.618 : 19.618. (See figure 6.7.)

In building *Heptaparaparshinokh,* we can therefore construct ratios that are its intervals, on a base where the hypotenuse of the triangle has been arced down to the base, to add the leimma of the eclipse year to solar year ratio (see figure 6.8). The natural length for the octave is therefore the solar year (that, most naturally, is 12 and 7/19 megalithic yards long). The Pythagorean leimma is then found to be 12/7 feet, a length

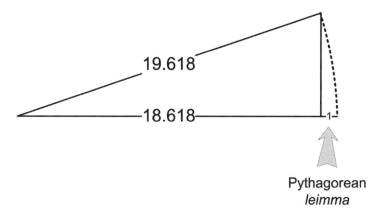

Figure 6.7. The eclipse–to–solar year triangle showing the Pythagorean leimma of 256/243.

known best from Egyptian times as the royal cubit. The justification for this approach is that the solar year is a recapitulation of creation in that the octave shown on the right of figure 6.9 begins in midsummer solstice and ends in the following winter solstice. In addition, this half year is built upon another octave that begins in spring, so that a quarter

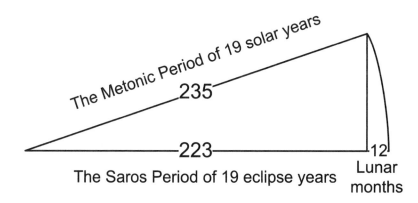

Figure 6.8. The Pythagorean leimma of the seventh Stopinder shown as the 19 eclipse–to–solar year triangle. The base is the well-known 18 solar year Saros period (between similar types of eclipse) and above is the 19 solar year Metonic period, after which the sun, moon, and stars have exactly the same positions relative to each other.

year makes another doubling of the octave, much like the vibrating string of the monochords used in acoustic experiments. One-eighth of a year defines the so-called quarter days, and there is evidence of a 16-month calendar in megalithic Britain—all powers of 2, or doubling.

The intuitive step of placing the leimma at the year-end changes the construction of a diatonic octave, putting 256/243 where the cosmic semitone 16/15 would normally be. Based on this approach, and by some similar logic, the changes of the sacred *Heptaparaparshinokh* replaces the third Stopinder with the lunar ratio 27/25, to balance the reduction by 81/80, of the seventh.

The 18.618 eclipse triangle is the larger triangle (drawn with dashed lines in figure 6.9), its base being compared to the diatonic scale since it ends at the end of the solar year.

The earlier semitone must absorb the shortening of the last semitone if the leimma is to replace it—the opposite of Gurdjieff's explanation. The notion of this substitution is suggested by the fact, previously mentioned, that the difference between the leimma and the 27/25 ratio (of the lunar month to orbit) lies in two syntonic commas. This means that these two lunar ratios replace the semitones without adjusting any of the other diatonic whole tones.

The 27/25 ratio then replaces the 16/15, first diatonic semitone, as the third Stopinder, which means that the end of the second Stopinder signifies the lunar orbital period, and the beginning of the fourth Stopinder is the lunar month. This is shown as the smaller triangle (drawn again with dashed lines; see figure 6.9).

The idea could have formed that the moon generates semitones that have a harmonic function and that the concept of a harmonic cosmology could emerge from such constructions. Especially noteworthy are the (unintended) similarities to the ray of creation in the diagram shown, as the notes do, re, mi, and so forth, formed as arcs.

Any triangle based upon the solar year simultaneously signifies the Metonic cycle of 19 years and the solar deity Apollo, who invented the lyre. The tradition that he visited a spherical temple in Britain during

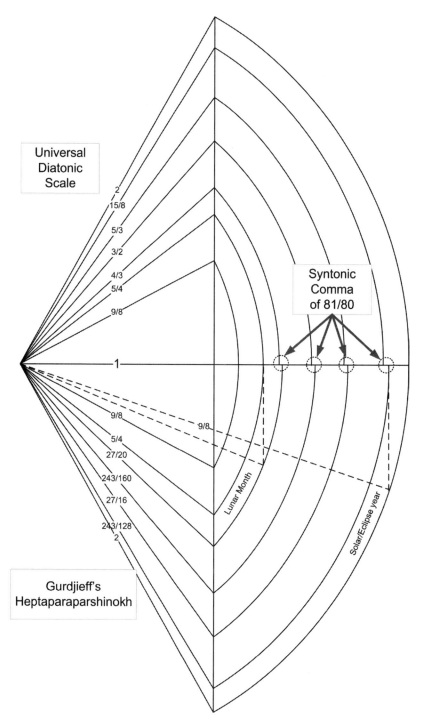

Figure 6.9. The geometric construction for comparing the diatonic octave and Heptaparaparshinokh. We are using the Megalithic procedures that employ metrological triangles as harmonic generators of known cosmic ratios. Within these, the necessary harmonic intervals exist.

the Megalithic period has been associated with Stonehenge or at least with the later Megalithic period. This is probably the time when the basic works based on astronomy were finished and the work on harmony and harmonic cosmology had begun. (Removing the need for the Megalithic Britons, Greeks, and Egyptians to have received these ideas from a preceding Atlantis.) The association of the lyre, an octave instrument, with Apollo—whose altar should be doubled, and who, it was told, visited Megalithic-period Britain every 19 years—supports the idea of a harmonic work having been conducted through metrology, triangles, and astronomical discoveries and available without any other mystery factors.

It is feasible that those who had harmonic cosmological ideas sensed that these did not work exactly as a diatonic octave, and therefore came to see the possibility that, through its presence, the moon (so close to life in every sense) adjusted the diatonic semitones. The result could then have been a teaching like that of Gurdjieff's sacred *Heptaparaparshinokh:* a harmonic universe, where octaves interact based upon a sophisticated awareness of interrelated processes.

A NEW TYPE OF INTELLIGENCE

The correlation between Gurdjieff's law of Heptaparaparshinokh and many of the results reported in my first book, *Matrix of Creation,* are unexpected. Ancient metrology has proved itself a tool for both astronomical and harmonic study that predates the work on musical harmony explained in chapter 3. Metrology was developed in the early Megalithic period, when the celestial time ratios were discovered and a fully consistent picture of creation was built of interacting vibrations. Instead of material causation and our modern notion of determinism, a higher level of cosmic causation must have been conceived in which the celestial bodies played a primary role.

The role of the celestial bodies would be seen as an enabling causation rather than a determining one of material causes and effects. The only familiar precedent for this is astrology, though modern astrologers often import the cultural concept of a deterministic causation.

An enabling causation would establish framework conditions that merely supply opportunities through environmental synchronicity and choices made, and therefore there is uncertainty as to *actual* outcomes. The third Stopinder relies on the mechano-coinciding of the law of seven in the environment. The ability to predict what would happen in the future was a major industry in early history, and augury and divinatory systems were designed to please the gods by cooperating with what they wanted to happen. These practices may have developed from efforts to read the enabling framework conditions.

The ability to further our goals relies on a number of appropriate steps as well as the availability of resources and motivations at the right times. The enabling of processes through the moon, sun, planets, and the structure of just seven processes would mean that situations are regulated by celestial ratios and harmonic laws, which can affect what to do and when and how to do it. In reality, the environment is a mix of diversity, chaos, and order. The arrival of resources, our decisions to attend to certain matters, the thoughts that are ours momentarily— these and many other factors could be subject to an ordering factor originating in the sky, and this is related to our idea that the precessional ages could be enabling cultural development.

In our world of chaotic inner and outer existence, there seems to be no special enabling factors or law of seven. We live in a world in which all achievements are simple victories of large-scale, superior forces, and this is the ruling political metaphor of *force majeure*. Problems are solved deterministically, statistically, and materially. The evidence is that this approach leads to catastrophic and unpredictable outcomes because the world is not linear. Linearity is simply a universal assumption in which determinism rules what happens, and this is what most human beings like to think. When linearity is imposed, it eventually breaks down in front of the natural order—having frustrated and hence intensified the natural forces that are then destructively released in a manmade disaster. The conscious mind has taken linearity and determinism and built upon these a logico-rational superstructure in which modern civilization has tried to replace the mantle of the planetary gods themselves, as an enabling causation but using deterministic means.

The first intelligence, born with the world and which we call the gods, maintains the created order. This intelligence is natural to the planetary bodies, and according to Gurdjieff, significant steps were necessary to ensure that the moon would not become a source of disharmony for the universe. We have seen that the moon is very delicately adjusted to fulfill the harmonic requirements of the octave.

A new application of the law of seven arises on planets with a moon such as ours—an adapted law that causes life to arise and then makes the conditions of life capable of generating a new type of intelligence. To develop this new intelligence requires the local law of seven to be applied; otherwise, the processes started will rarely bring about results or will bring about results that are the exact opposite of those intended—as is often the case in politics, for example.

THE PRECESSIONAL DANCE OF SUN AND MOON

WE HAVE SEEN THAT Gurdjieff associated all impulses to start things, and the environment that "reciprocally feeds" all processes, with our Moon. Since almost all of our impulses come from our culture and, because our culture feeds its own necessary processes, then a human culture appears to be manifesting the moon. Yet the moon is complemented by the sun as the main partnership behind precession, whereas the sun is higher to the moon's lower in Gurdjieff's ray of creation. The moon appears to define the first half of the octave, with the sun and what lies beyond it defining the second half.

As the achievements of a culture grow, there is an increase in the network of ideas and mutual supports it provides, becoming a nurturing moon, but only in that culture's name. Meanwhile, the descending energies of the sun for an Age are crystallized as the cultural mind that forms as the appropriately named note, sol.

The law of seven is a process of integration in which the story form, the latch of impression, is created by the sun but is manifested by the moon via synchronicity within the environment. In the working life of a culture, the sun and moon meet across the turn, where they meet opposite the root note, do. But this work is being undertaken by humanity through a largely unconscious response to cosmic influences.

The law of seven for human work requires us to make a start and attempt to progress with the help of culture, but once we encounter the solar half of the octave, we begin a contest with the cultural mind. Neither the sun nor the moon can substitute for what we have to do, even though these two were the makers of life in general and our own life in particular. We have a "natural" existence in which we are just part of the world machine and an essence that is more than living. (Bennett called this the *essence individuality* that is subconscious and leads to predispositions and synchronicities in life.) To realize our cosmic role, we must sacrifice some of our existence by taking risks and venturing into the unknown of creativity. We can realize ourselves only in the realm of the friction and clash of that which comes from above and that which comes from below. To become independent, we must cure the dependencies and assume responsibility for what is otherwise provided by nature through the culture.

It is not just impressions that "arrive" in time at do_1 of the octave, but also the higher pattern of human selfhood at conception. The story of our life is therefore a manifestation of that pattern as a sun, interacting through our bodily nature as a moon, working itself out within our culture.

Those who have discovered their path for individual development are rarely aware of what makes that development possible or of the technicalities of how it must be reconstructed, as opposed to simply being received. Intelligibility can be seen as an invitation to consciousness and to the task of reconstructing creative will within existence or bringing about its redemption.*

In Egyptian myth, this is the task of Horus, whose mother, Isis, collects the dispersed parts of his father Osiris's body. Osiris's body represents the cosmically creative energy. Set or Saturn, the usurping brother of Osiris, sees that Osiris is dismembered just as, in Greek myth, Kronos-Saturn castrates the Sky, Uranos, to create, from the genitals of Uranus, the planet Venus, who is also represented by Isis. Perhaps Isis can never find the genitals because they are the creation itself, which Saturn, as the outermost visible planet, circumscribes.

Every individual is born with a Pattern that impels them into face certain "difficulties" and the life experiences necessary to fulfill the creative energy of this pattern, as an autonomous being. With the creativity energy given us by the sun, our task is to rebuild the dispersed impressions of life into their original unitary nature as a further working of the cosmic Will.

---

*From *Websters:* **Redemption** 1. [noun] (Christianity) the act of delivering from sin or saving from evil. **Synonyms:** salvation 2. [noun] repayment of the principal amount of a debt or security at or before maturity (as when a corporation repurchases its own stock) 3. [noun] the act of purchasing back something previously sold **Synonyms:** repurchase, buyback.

CHAPTER 7

PRECESSION AS AN ENGINE OF CULTURAL CHANGE

THE PRECESSIONAL MYTHOLOGY OUTLINED in chapter 1 tells us there were some quite-advanced ideas about precession in the prehistoric period that employed a set of creative metaphors for its appearance and behavior. In the ensuing chapters, I have tried to paint a bigger picture in which our cultural history appears to evolve its consciousness within the dynamics of changing equinox signs. This reveals humanity's predicament in needing to take definite cultural steps that are orchestrated according to precession time that is itself organized in the image of a transformational cycle.

We can look back over half of this precession, assuming a starting point of twelve thousand years ago, to the end of the last ice age. This human journey first turned outward, toward an understanding of the visible structure of the universe. Then it turned to the structure of the number field that underlies creation. This set us on a path of technical and religious change that started the medium of written records—the recording of history. This outward-looking gaze found its terminus in our own scientific age, as physical sciences overwhelmed the idea of a God who organized and continues to organize the world directly. Technology, based upon this science, has led to the proliferation of devices and information that have a massive impact upon society today. Indeed, a single generation has now witnessed multiple socio-technological upheavals.

Within this recent period, the narrative of the culture is no longer shaped by the culture itself but by its technology and its media in particular. This means there is a problem of initiative, for it is tempting us to be passive in the belief that the future will bring ever greater technologically achieved happiness. Such passivity is not making conscious decisions but rather is dominated by the physical and emotional. The present human culture is falling into a state where the intellectual brain, emotions, and actions rarely act as a single whole, and those instances usually occur only by accident. This means that citizens act like two-brained or even one-brained beings, which is like thinking animals or even insects. There is therefore a considerable risk of the cultural moon overwhelming human nature altogether in this age of so-called rationality—where the culture is undermining the evolution of human consciousness.

If there are cosmic energies that humans can transform, then the human work of receiving and altering them is important. Again, Gurdjieff provides unique clues.

THE LIFE OF IMPRESSIONS

We have defined the three types of food for humans: food, air, and impressions, and impressions prove to be a very special kind of food. "Neither air nor food can be changed. But impressions, that is, the quality of the impressions possible to man, are not subject to any cosmic law."[1]

Impressions can come from different levels of the cosmos, and they appear to be something fundamental to the universe that makes itself known—a property we have called *intelligibility*. This allows for human consciousness, which can make available something in the lower worlds that can bridge the limitations of these worlds and thereby redeem them.

Impressions are not simply remembered experience, but, rather, *impressions are experiences that remember themselves,* that crystallize within our being as something that can be further digested. In Gurdjieff's concept of work on ourselves, the objective is to enable impressions to evolve into a cosmic will based upon understanding them. If humans receive impressions that carry higher vibrations—this amounts to "will doing being" rather than "being doing will." Ordinarily, we view ourselves as beings or sources of initiative who project our aims onto the world (our environment), rather as if we were creators. Instead, initiative involves impressions, and we can serve only their development within us. This development can project outward onto materialization of thought and inward onto self-development. In this view, we are co-creators with impressions. There is doing or action in which we can *participate,* and the higher worlds approach us via the law of seven, a process, rather than through agencies or beings, as religions have proposed.

Higher reality comes to us in a raw form, as impressions, and derives from a common source: a substance corresponding to the latent intelligibility of the universe. The common source substance, or reality, is will. This notion of will as coming from above corresponds to familiar religious ideas: that will originates in the higher worlds and the need for humans to respond to the will of God or the gods. Our ideas about saints, for example, include the story that they have been cleansed and therefore able to receive information from higher worlds. The need for a preparatory process corresponds to Gurdjieff's sacred sevenfold Heptaparaparshinokh, in its articulation of the will through cumulative actions and also to the compositional structure of storytelling.

As a procedure for doing, the law of seven takes an initial impulse and develops it. Impressions form the raw material of any impulse, which can come from any level through the medium of environment, codified here as the moon. Their further development can become blocked by the first semitone interval if we do not find in the environment something that corresponds to them and that can revivify them. In this way, a work will not proceed without transforming something in the environment to our own use. All of this involves the path of expansion for an impression within the world. In the second half of the octave—and in a story composition—the type of activity changes. It becomes that of perfecting and sacrificing all that is unnecessary in order to leave only what is then understood. This understanding is then the intelligible energy of the original impression transformed through a sacred action.

The last two Stopinders of the new law of seven are the means for realizing will, establishing the Trogoautoegocratic process upon which the universe was based. The new form of this law enabled the evolution of higher energies, seeded into the lower worlds as impressions, to return enriched by the necessary work that transforms those impressions. Those impressions that are untransformed are simply undigested and are recycled within nature; they have never truly incarnated. Beings who transform higher impressions are themselves transformed, but this cannot be achieved without some work on themselves, through which they learn to submit to the process.

MAN CANNOT DO WITHOUT INNOVATION

It becomes clearer why Gurdjieff said "everything happens, no one does anything."[2] Everyday, humans do things on every level, but we mistakenly identify ourselves as the source of those actions. Most of the superstructure of developing an ego and personality through an actual life within society ensures that the actions we take are personal. As a result, the idea of recognizing an impression as significant, fostering it until it develops a form, making an extra effort to perfect it, and allowing only the realization of what is essential—all this seems quite impossible to achieve. Yet we can achieve these, because they involve only operating with what is doable at each stage—not as saints or holy men, but people of the next stage in cultural development. In a sense, we must take from Gurdjieff what we can work with; not what is possible for the highest development of three-brained beings.

The history of humans is the story of doing what is possible and then finding the help of an invisible, serendipitous hand. Indeed, if our history were too miraculous it would also be broken history—for it would no longer be a path of development achieved by human beings. By introducing just a few ideas about impressions and process and combining them in our age (in which information and technology are almost fully realized), perhaps the life of our ordinary impressions can find a path of development that does not require the full rigors of spiritual work on ourselves.

If we cannot do or act from a place of self-sufficiency, then we can at least follow instructions, which is exactly what happens with the use of technology. In a sense, we have become slaves to machines, and today they develop us more than nature. It is likely that impressions arriving on Earth led to modern technology. Transformed into technology, these impressions have represented a hidden intervention in human life.

It would be useful to have a technique that can help transform impressions. Technology can play a necessary role in making easy what was once difficult as an externalization of our own development. A clairvoyant may not need a telephone, but a phone makes much of his or her capability available to those who need it.

THE ARISING OF FORM

The academics who study the compositional form of ancient documents are not themselves drawn to write using this form. The techniques of ancient storytelling have languished for hundreds of years, and there are few explicit examples of modern writers who use the old techniques of storytelling. Since narrative speaks of the forms within human life, this ancient technology can be used to understand impressions through the work of constructing a narrative form.

The idea of Form is deeply rooted in the Classical roots of our culture.

Plato's theory of Forms or theory of Ideas asserts that non-material abstract forms (or ideas) constitute the highest and most fundamental kind of reality, not the material world of change known to us through sensation. When used in this sense, the word *form* is often capitalized. Plato speaks of these entities only through the characters (primarily Socrates) of his dialogues who sometimes suggest that these Forms are the only true objects of study that can provide us with genuine knowledge. . . . Plato spoke of Forms in formulating a possible solution to the problem of universals.[3]

Plato's ideal Forms may seem similar to Gurdjieff's impressions, and Plato held a harmonic concept of how the world was structured similar to that of Gurdjieff. This sentiment led, however, to an idea of education in which impressions come purely from the forms found in our intellectual structuring of the world. This differs from Gurdjieff's idea of impressions being experiences of a three-centered (or "three-brained") human being. Plato's ideas, instead, tend toward the intellectual educational norm that has deeply influenced the modern world of the rational. While intelligible structures are abstract constructs, they are also *the experience of receiving them* through all three centers, and it is this that makes them real impressions.

The intelligible world we find in numbers cannot exist as we experience existence, because a higher world is less subject to existence and to the effect of time that mixes up intelligibility. Plato's Form is, therefore, intelligibility in the abstract. Instead, impressions are an everyday sense of a meaningful experience exactly because they are a communication between the cosmos and the biosphere taking place in us.

An impression relates to a moment of time in which there is a sense of latent meaning peculiar to the moment, of our selfhood and our environmental conditions.

Forms, on the other hand, may be

1. universal, such as number,
2. cosmic, as in astronomical objects and physics,
3. evolutionary, as in the species of evolved animal types,

4. descriptive, as in artifacts such as a table or a car,
5. social, as in behaviors and institutions.

All these types of form are unified by our abstracted consciousness, which becomes an internalized nexus of meaning, and this can buffer us from the meaning of real impressions.

Gurdjieff said the third food of impressions is our continuous receipt of the meaning of our world. He suggested that we would die in days without food and water, within minutes without air, and within moments without impressions.

Impressions are therefore sandwiched between the stream of personal experience and an abstract mind that holds the picture of what we believe the world to be. An impression as information cannot compete with the *formatory apparatus*—Gurdjieff's term for our abstract worldview—if it conflicts with such a strongly held view about the world. Though an impression's content consists of the same descriptive archetypes that we have learned to recognize within our abstract worldview, its vibrational energy generates an experience that is unusually memorable and meaningful. Such impressions can progress if they are recognized as significant and, only then, certain insights formed from them—though these cannot progress further without reinforcement from outside, the third Stopinder that must come from our environment.

To become food for us, these impressions must correlate with something new about the form of the world itself, and not simply reinforce the picture of the world already in our abstract mind. Impressions that transform into a new understanding can change the way we look at the world and the way we act.

Gurdjieff presented impressions as having three different sources. Here is a summary by Anthony Blake: "There are three main kinds of impression, which differ according to their source. Ordinary impressions come from the Earth and its surface. The second kind come from the Sun and the other planets, and the third from the ultimate source of reality Gurdjieff calls 'The Most Holy Sun Absolute.' Higher impressions can somehow 'lodge' in us but remain unassimilated, when they 'languish' inside us."[4]

From this perspective, major cultural achievements are not produced by our unaided efforts, and cultures themselves come from impressions received from higher, cosmic sources. If we can find a way to work with these cosmic sources, then these impressions have the power to bring about the appropriate changes to cultural life that relate to precessional time.

If the evolution of consciousness is driven entirely through the lives of individuals, and groups are able to work with higher impressions, no miracles are necessary. Free will is not compromised when we choose to work with higher influences and receive their implicit will to change the world. Centralizing organizations, such as states and religions, can support this only through maintaining conducive environments for the reception of higher influences—that is, through maintaining environments for their own cultural moon. Individual will, however, must have additional special environments to develop new and different ideas.

EVOLUTION FROM BELOW

The creation of higher energies from lower energies, as seen in the car engine, is the basis of most technologies. An internal combustion engine, through its design as an apparatus, turns heat (dispersed energy) into motion (directed energy). To do this, the higher energy found in fuels (a molecular or connected energy) explodes within a cylinder whose piston can move in only one direction. This apparatus, like living bodies, takes something higher from the environment (a fuel) to make something lower (the heated explosion) into a desired middle (a motive force), exactly according to Gurdjieff's law of triamazikamno or threefoldness.

The energy of the higher must be available, but it cannot increase order within a less ordered world without a suitable apparatus to make use of it. Forest fires are natural, but cars harness the energy of combustion according to our technical insight. Life was also based on this principle of necessary apparatus. This is the whole raison d'etre of the universe in Gurdjieff's view: the universe always strives to create apparatus within itself to generate something new.

The application of our intelligence to the task of interacting with cosmic energies relies on at least two key ideas similar to those present in the invention of the car. The first: to identify higher energies that can be assimilated. In the case of the car, this is a fuel formed millions of years ago by the biosphere, and in the case of human beings, it is the cosmic energy latent in our impressions. The second key idea: an apparatus constructed to transform cosmic energies into something greater than would otherwise exist.

Perhaps an apparatus could help cosmic energies to interact better with human consciousness if it avoided what normally happens to impressions. An apparatus for extracting higher energies from the environment would replace the wrong notion that we can already think about them without such an apparatus. The goal would be to generate actions informed by what is really meant by our impressions. Such actions would come naturally if we did not already think we knew what to do. A new technology might achieve this, instead of special training.

Before moving on to how we might better understand our impressions, it will be helpful to investigate how the internal combustion engine actually employs the law of seven. The internal combustion engine starts its octave with some momentum, which is usually available from the working engine but is initially provided by a starter motor or, in times past, by a starting handle. Do_1 is therefore this angular momentum.

Twelve semitone intervals are required to achieve the seven notes in an octave. In figure 7.1, four semitones are piston strokes and three are valve actions for exhaust and a delayed fuel intake before and after the carburetor that mediates the fuel and air from outside. All of the key semitones are in the fifth and sixth Stopinder, which concern explod-

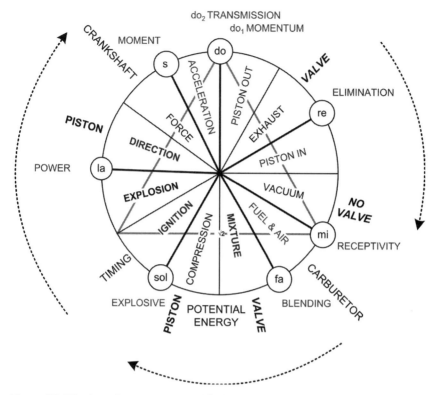

Figure 7.1. The law of seven as exemplified by the internal combustion (four-stroke) engine. The state of the engine becomes more sophisticated as the cycle proceeds. There are interesting oppositions and levels—such as force opposite the fuel mixture (which describes the basic idea), and the explosion is on the same level as the vacuum, which is conceptually an implosion.

ing a compressed fuel mixture and directing this explosion. The seventh Stopinder completes the aim of the engine in accelerating its crankshaft and providing the power of the explosion through a transmission system.

The central axis of the cycle has potential energy at the turn below and angular momentum at the latch above—a momentum that provides a perfect and necessary connection by keeping the engine turning and by making continuous the flow of power to the transmission.

In the absence of fuel, the first two Stopinders can be given energy to function as a vacuum pump, which is what they are. By adding the later Stopinders, we can draw into the later strokes a chemical energy to create a lesser energy of heat, which, in the form of a constrained explosion, can work to move the piston, and this work is transmitted through the crankshaft of the engine, which translates linear force into rotational acceleration and momentum.

TRANSFORMING ENERGIES THROUGH TASKS

Human experience operates using three energies as yet unknown by science. Much of what we do is automatic, and we can do things that we have learned with little or no further attention to how to do them. When we need to learn a new activity, we need a sensitive energy in which our attention is aware of exactly what is happening. Just as short-term memory progresses to long-term memory, so too the sensitive experience of learning is replaced by the automatic skill to act.

There would be no need for any higher energy if we did not need to understand, if we did not have a capacity to read what is intelligible about the world. This higher energy is associated with conscious as opposed to sensitive awareness, for which it is often mistaken.

Because we are potentially creative beings, we can touch upon a further energy that is harder to recognize, for only its results can be seen—when we see that the world has been changed. Therefore, this creative energy can only be seen as a reality by conscious energy that finds a creative solution where there was not one before. Many people can stand before a creative solution and not recognize that something special lies before them.

Only the lower two energies—automatism and sensitivity—belong to the living being, to Life. The other two—consciousness and creativity—arise as part of cosmic actions that are intelligible and creative. The border between the living and the cosmic is therefore found exactly where we find the two most significant energies to human activity—that is, the sensitive and the conscious. Therefore, any work we do is either sensitive or conscious. In a sensitive work, we can rearrange, and in a conscious work, we can understand what is there.

As humans we can be compared to an apparatus like the car engine: the sensitive and conscious energies are the only energies available for the working of such an engine, and so these activities must be organized. Yet what kind of a work seems most significant to transform one element into a higher element within an octave? According to Gurdjieff, we have the constant impression that we are alive, and humans are constantly called into action. Remembering the idea that the world is intelligible, it is this intelligibility that lies behind the world of impressions. As intelligent creatures, there are things we humans can do, based on the intelligibility of the world coming to us through impressions. Yet the impressions do not say, "Noah, build an ark." Instead, they might say, "Boat building summer school" as an advertisement and "Climatologist Heads for the Hills" as a headline.

There exists a system for doing based on impressions—a system that has been developed since Gurdjieff's time called Logo Visual Technology (LVT).* As with the car engine, what it does can be viewed within the structure of an octave, from impressions to actions.

The five processes of LVT—Focus, Gather, Organize, Integrate, and Realize—function as the five whole tone intervals of the octave, but each of these can be further split into two semitone tasks, one in which sensitivity is employed and the second in which consciousness must be employed in order to see what has been done using sensitivity. These can be shown as a table.

---

*Logo Visual Technology was developed by Anthony Blake from earlier ideas called Structural Communication (J. G. Bennett) and with magnetic hexagons and Communikit (Anthony Hodgson). The work was commissioned by John Varney for Magnotes (now LogoVisual) to rationalize and develop the processes used with magnetic media in the area of thinking skills. In later years the author worked with Blake and Varney, when ring composition became the mechanism of choice for integrating a complete exercise. The standard media for this work is: dry-wipe, three-inch hexagons that attach to magnetic whiteboards, available from www.logovisual.com, but the essential requirement is to have moveable pieces of text that can be grouped together.

Sensitive Task	Consciousness Task	Product
noticing	recognition	a real aim
gathering	articulation	molecules of meaning
clustering	distillation	domains of meaning
sequencing	integration	a new worldview
understanding	realization	informed actions

Working backward, we can say that almost everything we do today in the human world is based upon the worldview held by each one of us or as a group. Thus, there is already a strong path to action if we can create a new worldview. The sensitive task of understanding a worldview leads to the conscious task of realizing what that view calls us to do.

Working forward, from the start of this octave, we can note that the continuous stream of impressions is mixed and must be ordered. By noticing the impressions using sensitive energy, we must have a consciousness task that can recognize a significant boundary within which to concentrate our work. Gurdjieff called this task that of forming a real aim that becomes the seed that requires a new worldview to succeed.

Once such a real aim has arisen, we can begin the sensitive task of gathering insights from impressions with respect to that aim, which then act like a magnet for relevant impressions. Each insight must then be transformed by articulation, a task for consciousness, in order to create a *molecule of meaning*—a small but well-formed statement—from which the Logo part of LVT comes.

So far, LVT does not seem very different from brainstorming; the subject of a brainstorm is a question, problem, or topic, and the ideas produced by brainstorming are written down, like molecules of meaning. At this point, though, a brainstorm comes to a halt as a technique exactly because the third Stopinder is a semitone requiring an impulse from outside the process. LVT bridges this requirement by placing the written molecules of meaning on a set of identical physical objects so that together they create an externalized environment of independent meaning objects. These can then be moved around and connected to each other so that corresponding vibrations, contained within them, can bridge this semitone of the law of seven.

This connection of related meaning objects is a sensitive task called *clustering,* which leads to a set of grouped objects that have some potential meaning when we consider them together. The task for consciousness is then to find a name for that which each cluster holds in common—a task that is called a *distillation.* Each distillation statement infers the domain of meaning to which a cluster of meaning objects points, inference being an act involving conscious energy.

It is worth noting that conscious tasks are more difficult than sensitive tasks, because they create something definite out of an undefined sensitive awareness; they create the words of a real aim, molecules of meaning, and the naming of domains by using language in an accurate way.

The new domains of meaning are usually 1/3 or 1/4 of the number of meaning objects, which brings them into the range of episodes found in a ring composition of manageable size. At this stage it becomes relevant to connect the disciplines of ring composition to the fact that the stories produced by these techniques became influential worldviews. We are therefore growing close to connecting the impressions at do_1 with the actions at do_2 if the domains of meaning formed using LVT can be organized in the form of a narrative.

To achieve a narrative requires that the domains of meaning be sequenced, in a task for sensitivity. Though the domains are not a narrative in time, they can be given a pseudo-causal order, which makes some sort of a story out of them. If we can see any given domain as flowing naturally from another or as preceding it, like a set of dominos having compatible ends, then a single pathway through the domains of meaning can be forged and each transition between domains merely needs to be plausible.

Once we create a narrative, we can note a beginning, an ending, and a middle, which will create the latch and turn of a ring composition as soon as the domains are re-displayed in a cyclic fashion. The parallelism between left and right will also emerge. As with the explosion within a car engine, the second semitone of this, the fifth Stopinder, is unusual. In this case, the task of consciousness is formed by the act of displaying a narrative in cyclic form, which naturally generates the new and powerful insights found within a worldview (see figure 7.2).

The task for sensitivity within a ring composition is to play with the different perspectives it provides in order to develop some understanding of what these perspectives mean. At any time there will arise moments of consciousness that will realize what the constructed worldview reveals *that can be done* (see figure 7.3). If we have an authentic relationship to our subject, then a will can emerge that has its roots in our do_1 of impressions but has been transformed by the intentionally-actualizing-Mdnel-In into a do_2 of creative actions.

A WORLDVIEW FOR PRECESSION

It is clear that the modern world has been accelerating, which means that time itself has become subjectively faster—largely because there is a great deal going on that drives human events. But it is technology that has increased the pace of change, and each of the past cultures contributed to this acceleration. We have already proposed that cultures will naturally cause such an acceleration and that, while the moon orbits Earth, our satellite's realm of manifestation is actually in the biosphere and the cultures that ride on top of the biosphere.

If we apply the harmonic model to this cultural change, then we discover that new notes are produced by the intervals of precessional time associated with major cultural themes. A new note is a higher vibration, and, seen from the perspective of time, it is an acceleration of the subjective flow of time produced by cultures that provide, within themselves, new and easier ways of doing things. This means that while

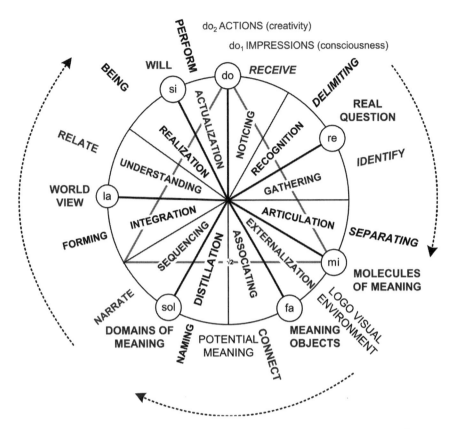

Figure 7.2. The process of turning impressions into creative action according to the medium of Logo Visual Technology (LVT). The five Stopinders of activity have sensitive and conscious tasks, each worth a semitone. The third Stopinder is facilitated by combining text with moveable objects to build a logo-visual environment in which the meaning objects can reciprocally relate to each other in the task of clustering. Through the consciousness task of distillation, the domains of meaning are few enough to be sequenced as a narrative and so made into a worldview using ring composition. The well-known power of worldviews to evoke actions based upon such worldviews can then complete the octave, because the intentionally-actualized Mdnel-In is derived from the worldview and the activity of generating it.

precession is a story in twelve chapters, it is also an octave in which the quickening of consciousness is caused by the cumulative effect of cultural development.

Success creates growth and acceleration, and when cultures fail they

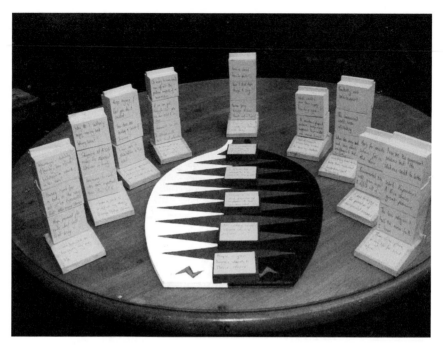

Figure 7.3. An experimental medium for enabling insights based upon impressions (our life experiences) to be transformed into a motivational ring composition. The viewer is in the position of the latch and is opposite the turn. The story begins on the left and ends on the right, and levels of meaning run up the axis of the story, mediating left and right. These levels are used to realize what we want to do based upon this narrative as a worldview.

leave room for the growth of new cultures. For such cultures, this growth is from a higher point and includes skills that have already been won and are not themselves destroyed with the original culture that brought them forth. Great civilizations have passed along a surprising number their important gains to later civilizations, but they must wane themselves as a direct cultural influence if there are to be significantly new cultures to follow. The harmonic model, therefore, fits with today's experience of extreme acceleration and of the last twelve thousand years as a gradual build up in vibration where the cultural tempo appears to grow faster.

For these reasons, the law of seven might apply to the precessional cycle. Given the previous analysis of how an apparatus raises energy to follow the law of seven, then we can see that the twelve precessional ages are naturally semitones that raise the vibratory level of cultural intensity, as in an apparatus.

The third Stopinder is the Age of Aries (see figure 7.4) in which written history and numerical notation created a great expansion of the cultural moon. We may compare this to the intake of fuel in an engine and particularly with the Logo Visual environment (of the previous section) in which meaningful objects are created to bridge the third Stopinder. It appears that the acceleration of consciousness is a function of the amount of cultural knowledge in circulation. The expansion of printing, literacy, and other mechanisms has each created an acceleration.

Writing and numerical notation were each a spin-off of the

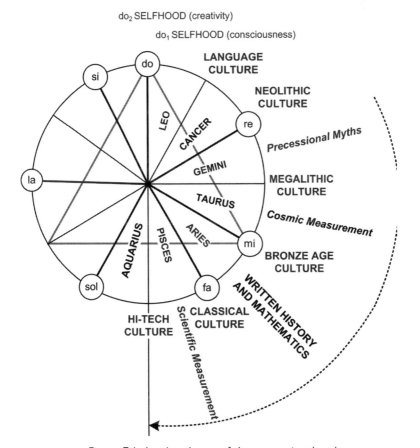

Figure 7.4. A cultural map of the precessional cycle

Megalithic period that empowered the next semitone to be bridged through the formation of an extensive cultural environment in which vibrations could combine. The result was a consolidation of religious symbolism and texts designed through harmonic realities. Out of a pantheistic cloud there condensed a monotheistic God who was involved in a battle between good and evil that would be broadcast to the world.

If there are energies of consciousness and creativity that are cosmic, that are not generated by life itself, then these pass through intelligent life as part of Gurdjieff's ray of creation. The modern condition has been caused in part by our thinking that we are naturally conscious and that we are already creative in the cosmic sense. If we are actually conscious or creative according to cosmic impulses that work through us, then the exact opposite of what we think is happening. This is perhaps the best way to explain why there is such a great problem and opportunity today in recognizing that consciousness and creativity are gifts as opposed to being naturally inherent.

THE SCIENCE OF STRUCTURAL WORK

Gurdjieff's law of seven is significantly different from the science of musical harmony in that the law of seven attributes notes as higher and stable levels of *structure*. The intervals between these notes are therefore the types of work required to transform one stable structure into another. This does not correlate with the experience of musical tones, because their structure is simply a waveform with a given frequency. It seems, however, that the intervals between structural transformations follow the same laws of harmony: the notes of greatest harmony occuring within the model of the major scale.

Structure can be one of two types when it is developed. There is extensive structuring in which a series of components are generated in preparation for a later intensive structuring. In this second phase of structuring, the parts generated in the first phase are assembled to achieve a new stable form, equivalent to achieving a musical note. In a story cycle, the first half is largely development, and the second half largely welds these developments together in such a way as to achieve closure for the story as a whole. All of our cultural artifacts and creations follow this pattern; it is the way in which higher results are produced from existing components in the cultural environment.

When we look at figure 7.4, it may seem a mystery that the Neolithic culture required the Megalithic culture in order to arrive at Bronze Age culture. But, by externalizing the principle of number and cosmic observation, many of the skills involved in making Megalithic structures and measuring astronomical time were later needed for metalworking and for numbers used in trading. Early written language retained a correlation between its alphabetic characters and numbers, revealing that the extensive Megalithic period work regarding number led to subsequent intensive work of studying the laws of number, geometry, music, and astronomy as disciplines—still relevant in Medieval times as the Traditional Arts.

The stories prior to the Megalithic period—stories from the Neolithic period—must have been intensive, because there was no other approach than to subdivide the sky as a *whole* phenomenon rather than to extensively count time: Megalithic people had to do extensive building work on aspects of this sky phenomenon, revealing time's numerical structure. This revealed the number field itself as an intensive pattern involved in the cosmic creation.

Structure differentiates types of materiality and gives rise to recognizable identities of forms that, like notes, are stable in their own right. A comet, a brain, and intestines share the characteristic of having an extra fractional dimension; they are all more than simply three-dimensional, a difference called complexity. New capabilities arise with fractal dimensionality because of their increased intensive structure. Sometimes, the surface area increases and at other times, it is the number of connections between parts that allows new behaviors to arise for the system as a whole.

The inner connectedness of a worldview, shown as a ring

composition, is an externalized statement of the connectedness of the world. This makes it perform better than simply talking, thinking, or writing linear texts. An extra fractal dimension appears with the inner connections and those who wrote ring compositions employed this extra dimensionality. It appears that we have a structural side to our being perhaps because we have two eyes, two sides, and an up and a down with which we work at making meaning.

When we draw a circle we present a view that, when populated, becomes a world. This world is a blank piece of paper for the side of us that lives on a sensitive screen upon which our world paints itself. A painting is a work of an experience and not of what is present in reality. An experience is a structuring of reality and is not reality itself. Some may have higher experiences when others in the same situation do not.

We are natural structures of meaning and only became scientific recorders of the exact nature of the universe by sacrificing ourselves to the factual. The movement into structure reveals a harmonic nature to experience that can evolve through extensive and intensive work so that new structures of meaning can be achieved. Because we did not create the universe, we can do things only by working on structures in the appropriate way. Any culture that succeeds ours may recognize the implicit structure of situations and develop ways to communicate using a developed structural science.

Epilogue

THE GENIUS OF CREATION

ONE IMPORTANT WAY TO encourage significant impressions is to go on a pilgrimage. Thirty years ago I visited a temple complex and hill associated with Parashurama (one of the avatars of Vishnu) in the Western Ghats of India. After a few days, the Brahmin families, with whom we had friendly relations through an English speaker, began to build a fire altar before entering into a meticulous recitation that continued for hours. They performed beside an algae-stricken bathing ghat whose surface was like a glassy mirror. In the sonorous and monotonous background of the smoky chanting, I could see distant palms reflected in the mirrorlike surface of the water as if looking into another world. Something in me clicked, and the experience became permanent, though I did not fully understand it.

I knew nothing of the splendid nature of the Agni altar ceremony, the number of its bricks and its syllables, and so forth. The conditions did not seem to have created the experience, yet my impression was perfectly appropriate to the philosophy of the *Tripura Rahasya,* associated with Parashurama, as became clear years later upon reading this book. In India, Tripura is a female Trinity. She upholds the whole of existence through a kind of illumination "which shines self-sufficient, by itself, everywhere and at all times."[1] The book's teachings were purportedly first given to Parashurama by Dattatreya* as found in *The Mysteries beyond the Trinity*.[2] One of the most significant metaphors used within the book was that of the reflection found in a mirror:

> 46. She holds everything by Her prowess as a mirror does its images. She is the illuminant in relation to the illumined.
>
> 47–49. The object is sunk in illumination like the image of a city in a mirror. Just as the city is not apart from the mirror, so also the universe is part and parcel of the clear, smooth, compact and one mirror, so also the universe is part and parcel of the perfect, solid and unitary consciousness, namely the Self.
>
> 52. Just as a mirror, though dense and impenetrable, contains the image, so also pure consciousness is dense and impenetrable and yet displays the universe by virtue of its self-sufficiency.[3]

Such an analogy is easily discounted as an over simplified conceit but it has a relationship to ideas in Gurdjieff's thinking; for the emanation of the Absolute, the *Theomertmalogos,* has a similar relationship to such "illumination," relative to the universal creative substance *Etherokrilno* that effectively forms a mirror for the intelligible nature of the world. Gurdjieff's vision however goes beyond the mirror analogy of non-dualism (found throughout non-dualistic Yoga) to reveal the missing inner texture of the resulting Creation or Trogoautoegocrat, as is perfectly put by Bennett in his *Talks on Beelzebub's Tales*.

> . . . in the unpublished version [the trogoautoegocrat] was indicated by the word *fagologiria,* the reversal of forces by eating and being eaten. . . .

---

*The ancient spiritual teacher or god who embodied the three-foldness of the creation by the having three heads shown in his iconography, those of Brahma, Vishnu, and Shiva (creator, maintainer, and destroyer gods). The three-fold nature of Tripura suggests her as his consort.

The trogoautoegocrat was made possible, Gurdjieff says, by changes in the functioning of the law of threefoldness. In its original form, the law was simply a successiveness; one thing arises out of another and passes into a third. Then it was altered so that something new could arise. This can be roughly shown as:

There is now a difference in the role of the three forces. There is a place where two paths meet (C), one place where a path goes into a succeeding one (B) and one place where it all starts (A). This gives an end point, a transitional point and an initiating point.

It is this change of working that enables something to be concentrated at a point.

The initial impulse coming from the Sun Absolute, the theomertmalogos, is destined to produce an independent creation and this must have its own spiritual life or freedom and is therefore the carrier of the [reconciling] third force. From the very start there is an invoking or foreseeing of the end to be reached.

To arrive at this creation, the creative act has to pass through a receptive or denying force. This gives the form of action symbolized by the formula 1-2-3: affirming-denying-reconciling.

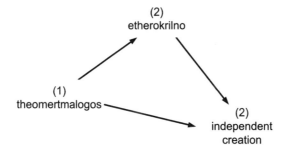

We can say that from the start one is looking towards the end as
an idea and one is moving towards the end as a fact.[4]

The etherokrilno is effectively the dense mirror within which the
universe appears. The missing element in the *Tripura Rahasya* is the
entire purpose of the creation—other than to form an illusion within
which the self is lost in ignorance of its projection out of the Godhead.
The purpose is to create something new, something independent that
has merit in its own right. Simply to return to the godhead is not indi-
cated. Gurdjieff proposes that new, independent intelligences called *cosmic individuals* are a possible and a necessary evolution of this new form
of creation.

The primary difficulty in our modern consciousness is its denial that
the universe has elements within it to suit our form of experience or that
intelligence is inherent within the intelligible nature of the universe. Yet
we have seen that the universe does have powerful organizing principles
regarding number and form and that these have been highly influential
in the creation of our past and present cultural moons. Intelligence and
intelligibility appear to be two sides of the same creative impulse—one
moves toward an intelligible universe as a fact and the other moves toward
higher intelligences as an idea for an independent creation.

The twin streams of cosmic emanation, shown to left and right on
page 177, hold between them the evolution of life as coming from the
substantial nature of the sun-moon-earth and progressing toward the
idea of cosmic individuality through the evolution of living consciousness. Life can be seen as being the translation of intelligibility into creative intelligence.

One suspects this simple truth was easily lost in the rubric of the
non-dualists of the Advaita and Buddhist Schools in their project of
returning to the self. Gurdjieff's vision had an additional and sane ethic
that did not reject the creation. The revised rule of interdependence
within the creation makes an independent creation interdependent
on itself. Rationality today prefers the prior form of Gurdjieff's law of

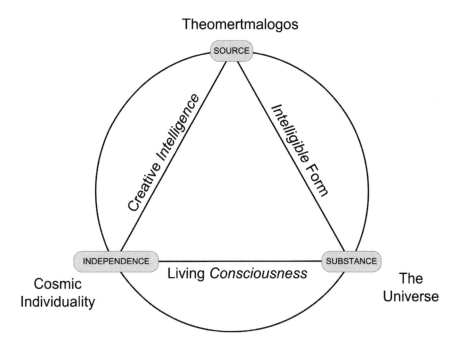

The emanation of the godhead forming the drama of consciousness and the universe. Implicit in this are the ideas of harmonic development and of the story form. Doubling is the evolution of an independent creation based upon consciousness and transformed through creativity, realized first by cosmic concentrations and then by independent, living beings.

threefoldness in which progress is in a line as successive cause and effect but then, time will inevitable shrink the rationalist's world just as our creator saw that time was shrinking his Sun Absolute:

> It was just during this same period of the flow of time that there came to our CREATOR ALL-MAINTAINER the forced need to create our present existing 'Megalocosmos,' i.e., our World. . . . From the third most sacred canticle of our cherubim and seraphim, we were worthy of learning that our CREATOR OMNIPOTENT once ascertained that this same Sun Absolute, on which HE dwelt with HIS cherubim and seraphim was, although almost imperceptibly yet nevertheless gradually, diminishing in volume. . . . As the fact ascertained by HIM appeared to HIM very serious, HE then

decided immediately to review all the laws which maintained the existence of that, then still sole, cosmic concentration.[5]

Here Gurdjieff presents a simple universal principle that applies to us as much as it does to the creator and applies to the place of our own existence as much as it does to the sun absolute. Though it might be interesting to know what was going through the creator's mind, it is more important to understand that similar thoughts need to be going through ours about how we try to achieve things.

Rationality is a tool alongside others including the structural skills of the ancient cosmologists and storytellers. The intelligibility of the world has a source in our own possible intelligence, which is not purely rational. The change Gurdjieff reported in the design of the universe is an indication that individuals must find for themselves a new way of working in order to achieve an independent creation—which is the reason for Gurdjieff's presentation. The prior law of successiveness is really a law of one thing after another and we can assume that the changed law of seven required the new law of threefoldness to be established in ourselves: that need to create something that will not just run down and then disappear forever.

REFLECTIONS ON THE BIG BANG

The will of Gurdjieff's creator, as the note do_2, descends into the harmonic universe as the note si (the theomertmalogos), and the substantial nature of the universe emanates from the lower do_1 to begin creation's work in acting upon the creative ether (the etherokrilno). These actions from high and low do are symmetrical semitones. We know that the early universe was a corresponding big bang of pure energy with very little structure but that the early universe very rapidly cooled into particles and then into the prototypical atoms we know as hydrogen and helium, the lightest elements. This cooling caused points of concentration of matter that, within galaxies of such matter, started to condense as stars. Stars in general and our sun in particular form the cosmic note re after the first Stopinder (as in figure on page 183).

The second Stopinder of the universe is therefore concerned with worlds caused by condensation, initially condensation into stars that created all of the heavier elements, only later to form a generation of stars made up of multiple elements, such as our sun. Gurdjieff called stars secondary suns (*deuterocosmoses*) because they resembled the sun absolute (the prime concentration or unity) after the universe was created. Just as the etherokrilno is a dynamic subnote within the first Stopinder, a solar system for our sun is a subnote in the second Stopinder. This solar system is a dynamic medium whose nature manifests in the numeric properties of the orbits of the planets. Our solar system is harmonically organized, especially when seen from Earth, one of the third-order suns (*tritocosmoses*) sounding the cosmic note mi.

The formation of the planets completes the material creation that we experience, as four different types of energy come into play in the first four successive semitones. The Big Bang was pure energy without structure, a *dispersed energy* that we now experience as heat. Electromagnetic and gravitational fields developed the structure of the universe as galaxies filled with stars, and this *directed energy* is a higher, more structured form of energy, with gravity forming the solar and planetary concentrations. The suns produce heavier elements through thermonuclear fusion to create a fully chemical world, and this chemical or *connected energy* is again a higher energy that can condense these elements into planets. The volcanism, gravitational tides of moons, and viscous geology of planets creates a higher *plastic energy* in planets with molten cores to create what, on Earth, is a living planet with cycles of subduction, carbon, and water, amongst others. Yet at this note of mi, all this can come to a halt if there is not a large moon present such as ours, created by "something corresponding, from outside" at this third Stopinder.

To go further requires a chemical energy that is higher than mere connectivity or plasticity. This is a *constructive energy* in which chemical structures evolve, replicating their own form as a basis of life. These molecules are long-chain proteins or nucleic acids such as those found in the DNA and RNA required for evolution. The ability to replicate and maintain structure makes a new type of creation that is autonomic, for living structures are able to regulate themselves by feeding upon their environment. Earth achieved this third Stopinder in its collision with Thea to form the moon and thereby to create a strongly interactive satellite. The moon kept Earth geologically alive while Earth established conditions for replicating molecules to develop into independent unicellular life (*microcosmoses*), which was to form the turn, opposite the Big Bang of the sun absolute—and become the central meaning in the creation. Life is a source of new arisings that are possible only through the first half of this cosmological story cycle and its energies. As with the etherokrilno and the solar system in the earlier Stopinders, life is a dynamic medium that leads to the cosmic note sol—the ordered environments within the biosphere of life.

Our cellular life early on required the internalization of mitochondrial DNA to transform oxygen into *vital energy* without any oxidation of our living material. At the same time, the primary mechanism for building cellular bodies was photosynthesis. This process builds carbohydrates from the raw materials found in water, atmospheric carbon dioxide, and the higher energy found in sunlight. Once we have assimilated carbon in this active and useful form, we can burn or oxidize it by the vital processes found in cellular life. This vital energy is higher than constructive energy and is the secret behind life. It leads to the automatic energy of the animal kingdom. Energy from carbohydrates and the oxygen captured by cells achieves actions independent of the universe, in animals that can move.

And thereupon, when our COMMON FATHER ENDLESSNESS ascertained this automatic moving of theirs, there then arose for the first time in HIM the Divine Idea of making use of it as a help for HIMSELF in the administration of the enlarging World.[6]

The highest manifestations of the note sol are our human cultures. These can be attributed to the precessional cycle, which lasts nearly twenty-six thousand years and is largely defined by the moon. Human cultures are largely automatic and unquestioning in their functioning, dominated by the goals of society. Yet there are additional sources of creativity within cultures that innovate them. A culture is between the two higher energies of life—a lower, *automatic energy* that burns oxygen in action and a *sensitive energy* that forms the types of experiences a culture chooses to pursue. The vital and connective energies are below our cultural awareness, as these are carried out for us by our bodies and are taken for granted.

The descending energy of the creator's will, which initiated the creation, becomes relevant only once intelligent beings come into existence. The arising of life at the turn begins this movement into intelligence within the second half of the cycle, which belongs to the inner energies of the outer, cosmic concentrations. In the fifth Stopinder, beyond culture, lies the next subnote of our human intelligence as it begins to interact with the cosmic world and goes beyond human culture. Human intelligence exists at the interface between the sensitivity of ordinary existence and the cosmic energy of consciousness, which has to do with the intelligibility of the world. This intelligence can form a direct connection, through consciousness, with the cosmic agencies of creativity—the angelic beings or demiurge that embody the creative intelligence of the universe. Such higher beings correspond to the sun and solar system, and they are sometimes called the heavenly host.

Human creativity feeds human culture with new artifacts of human intelligence by receiving a *creative energy* that is higher than consciousness. This creativity can be experienced only by recognizing a creative step when it occurs. Mere sensitivity cannot receive such creativity, because this creativity cannot be recognized without the capability and wish to understand it. Consciousness is transpersonal, because it arises only in connection to understanding. Creative energy is the continuing creation of the world and can be made intelligible only by *conscious energy,* which is the energy of the intelligible. This is a very great miracle; past, tentative steps to understanding the world have prepared us to be able to receive directly its creativity. The work of older cultures in understanding the outer universe has developed intelligibility so that, on occasion, we can face the creative process as an immediate action.

This potential transformation of humanity is the truth behind the rumors of human transformation into a form of perfected beings for the continuous work of the creation. The achievement would then be of the higher do_2, an expanded universe that expresses the divine without compulsion. Through these means, the creator has reproduced—created— another "like unto himself" (through the re-creation within ourselves of a similar creation). Such creation goes beyond the creative energy and reveals a unifying energy that might be called love and which maintains the creation's unity through a common bond that all beings must have, as with the interdependences of the trogoautoegocratic process. Even the reciprocity of eating and being eaten, found at the third Stopinder, expresses this *unitive energy*—as does the harmonic cosmological cycle as a whole.

To achieve all this, the creator must first have transcended or renounced the universe in the sense of only enabling the conditions for new arisings, and not directly controlling them. The revised sun absolute expresses a *transcendent energy* between the notes si and do in the semitone of that sun's expansion, which forms Gurdjieff's intentionally-actualized-Mdnel-In and completes the octave with a doubling of vibration.

The right and left sides of the cycle complement each other; they are the outer forms and inner beings of creation. The bottom of the cycle also reflects the initiating do in that the new arisings have an ascending character that complements the descending nature of the cosmic beings associated with cosmic concentrations. The path of ascension forms in the intermediate semitones within the cosmic octaves, for example, as

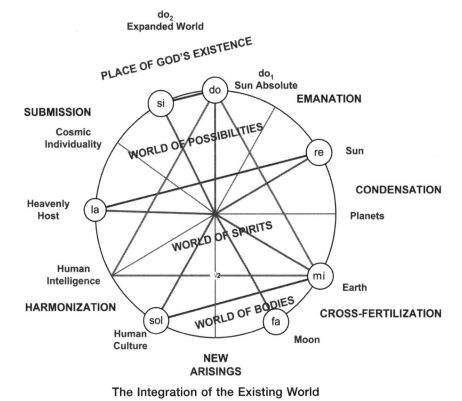

The Integration of the Existing World

The form of the creation and all within it. "What is very difficult to get hold of is that these other worlds are not somewhere 'over there' across some kind of spiritual Atlantic. They are here and now—in fact, more really here and now than our ordinary experience allows us to be."[7]

the human intelligence held between the sensitive and conscious energies and the cosmic individuality held between the creative and unitive energies. Born as breech babies (at the square root of 2), we are creatures of the in-between subnotes, which are dynamic and in flux, though the cosmic notes are established and rooted in the outer cosmic nature. Between conscious and creative energy lies the heavenly host (the note la) from which creative acts can be received or recognized by consciousness for what they are or could be.

The precessional cycle operates in the bottom third of the cycle and reflects God through the mechanism of earth, moon, and sun. This provides a cultural platform (the note sol) for human intelligence and its possible transformation. The distance from God is great only in the outer sense of cosmic immensity, because every process operates on the same basis and includes aspects of everything in its completion. This is why mysticism has in general concerned itself with the balance between the inner and outer worlds of humans and receptivity to higher worlds and energies. As the *Gospel of Thomas* says, "But the Kingdom of the Father is spread upon the Earth, and men do not see it."[8]

A final step in this story of how stories relate to consciousness appears in the traditional notion of a higher self associated with human beings. It is said that each individual has a higher mind of opposite gender—the animus for women and anima for men—or Sophia, Greek for higher intelligence. This self, as with the dream storyteller, can tell stories of *meaning through life* as a medium of communication to the selfhood located within existence.

The virtue of such a higher self is that it is free from the outwardness of existence and can reveal an inner life in which meaning lies between the events and structures found in existential situations. This aspect of reality, that we as humans perceive as meaning, is left "open" by an existence that only determines factual reality, so that meaning can still subsist within an existence whose only other meaning is its causal necessities and their effects.

Just as the notion of a Creator is natural to explain the highly constructive order found in the universe, so also the higher self is an equivalent creator for the selfhood and the high degree of order and structure found in the psyche and its life events. Such a personal creator of oneself is an "imaginal guest" thought to be essential to the development of the soul and the working of consciousness.*[9] Seen

---

*For instance, from Mary Watkins: "Once we open up reality to include the poetic, the dramatic, and the spiritual, the development of our relations with imaginal figures can no longer be confined to our customary notions. Development itself needs to be reconceived. Adaptation to reality changes its meaning, as reality becomes not just the sensible, material, and external reality, but created and imaginal realities as well."

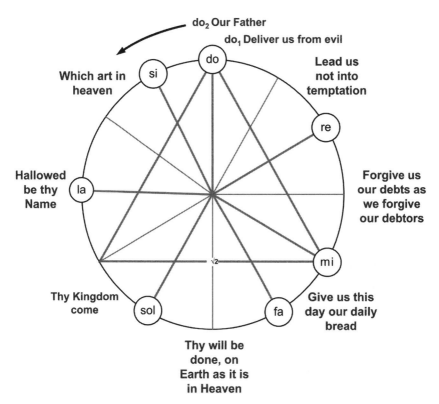

The Lord's Prayer as a descending ring composition. The prayer has nine main statements. The first and last form the latch: "Our Father" and "deliver us from evil." The turn is the central meaning: "thy will be done, on Earth as it is in heaven." As it is spoken, the prayer descends through a narrative of power and glory and then provides the path to this, in what is shown on the right in the figure.

This prayer is part of the Gospel of Matthew—it comes not from Mark but from a Q document of the sayings of Jesus, and we can see that the practice of ring composition resembles poetry in its ability to deliver a structural message within a few lines of text. All of the cosmology and harmonic meaning lie beneath what appears to be a compact formulation with no particular structure, but everything is in the right order and in its right place.

from existence such an imaginary totem is a fantasy but the arising of something unconditioned within existence would have to employ a faculty of spiritual structuring as that found in the imagining of such a higher self and a greater life story for the individual. When there is no meaning in life, consciousness has effectively departed, but meaningful life stories always contain the notion of an implicit higher self or animating principle.

Precession itself probably holds open such connections to the world before existence. This would manifest in the exuberance of what is possible in each of its ages but also as the melancholia of what is no longer possible, once an age is ending.

NOTES

CHAPTER 1
LANGUAGE AND THE
CONSCIOUSNESS WITHIN STORIES

1. For more on value, see "The Two Domains" in Bennett, *The Dramatic Universe,* chapter 25.
2. Douglas, *Thinking in Circles.*
3. For more on fact and value, see "The Two Domains" in Bennett, *The Dramatic Universe,* chapter 25.
4. Tilak, *The Arctic Home in the Vedas* and *The Orion.*
5. de Santillana and von Dechend, *Hamlet's Mill.*
6. Revelation 21.
7. Sullivan, *The Secret of the Incas.*

INTERMEZZO
THE GALAXY MARKS TIME

1. Shri Yukteshwar, *The Holy Science,* xi.

CHAPTER 2
MEASURING THE COSMOS IN TAURUS

1. Heath, "The Discovery of a Soli-Lunar Calendar Device within an Astronomical Ritual Complex at Le Manio, Morbihan, Brittany."
2. Thom and Thom, *Megalithic Remains in Britain and Brittany*, 90.
3. Robin Heath, *Sun, Moon and Stonehenge,* chapter 3; Hoyle, *On Stonehenge,* chapter 3.
4. Robin Heath presents this discovery in *Sun, Moon, Earth.*
5. Heath, *Sacred Number and the Origins of Civilization,* 198–202.

CHAPTER 3

GOD AS HARMONY IN ARIES

1. Ernest McClain, "The Tonal Calendar," 33.
2. Ibid., 101.
3. Rig Veda 1.164.48.
4. See www.bibal.net, Yahoo Groups forum, McClain.

CHAPTER 4

CRISIS OF THE MIND MACHINE IN PISCES

1. See http://en.wikipedia.org/wiki/The_Sorcerer%27s_Apprentice (accessed March 9, 2011).

CHAPTER 5

GURDJIEFF'S LAW OF SEVEN

1. Ouspensky, *In Search of the Miraculous,* 58.
2. Ibid., 124.
3. Ibid., 122–23.
4. Ibid.
5. Ibid., 124.
6. Ibid.
7. Ibid., 125.
8. Ibid., 129–30.
9. Ibid., 132.
10. Ibid., 139.
11. For information related to DNA and the golden mean, see http://goldennumber.net/dna.htm (accessed October 6, 2010).
12. For example, see figure 8 in Michell, *The Dimensions of Paradise,* 35.
13. Gurdjieff, *All and Everything: Beelezebub's Tales to His Grandson,* 753–4.
14. Ibid.
15. Ibid., 900–901.
16. Ibid., 84, 86.
17. Ibid., 770–71.
18. Ibid., 410.
19. Ibid., 754.
20. Ibid.
21. Ibid., 753.
22. Ibid., 754
23. Ibid.

INTERMEZZO
GURDJIEFF AS AVATAR

1. Bennett, *Gurdjieff: Making a New World*, 82.
2. Ibid., 189.
3. Gurdjieff, *Beelzebub's Tales to His Grandson*, 1094.
4. Bennett, *Gurdjieff: Making a New World*, 185.

CHAPTER 6
MULTICULTURALISM AND THE END OF
TRADITIONAL KNOWLEDGE

1. Gurdjieff, *The Herald of Coming Good*, 26.
2. Gurdjieff, *Beelzebub's Tales to His Grandson*, 753.
3. Ouspensky, *In Search of the Miraculous*, 85.
4. Schwaller de Lubicz, *The Temple of Man*, 250, 253.
5. Rowlands, *Zero to Infinity*, 101.
6. Bennett, *Talks on Beelzebub's Tales*, 98.
7. Gurdjieff, *Beelzebub's Tales to His Grandson*, 754.
8. Bennett, *Talks on Beelzebub's Tales*, 99.

CHAPTER 7
PRECESSION AS AN ENGINE OF CULTURAL CHANGE

1. Ouspensky, *In Search of the Miraculous*, 321.
2. Ibid., 22.
3. "Theory of Forms," http://en.wikidpedia.org (accessed October 8, 2010).
4. Blake, *A Gymnasium of Beliefs in Higher Intelligence*.

EPILOGUE
THE GENIUS OF CREATION

1. Haritayana, *Tripura Rahasya or The Mysteries beyond the Trinity*, chapter 14.
2. Haritayana, *Tripura Rahasya*, chapter 14.
3. Ibid.
4. Bennett, *Talks on Beelzebub's Tales*, 63–64.
5. Gurdjieff, *Beelzebub's Tales to His Grandson*, 749.
6. Ibid., 762.
7. Bennett, *Deeper Man*, 217.
8. Guillaumont et al., trans., *The Gospel According to Thomas*, Log.113, p. 57.
9. Watkins, *Invisible Guests: The Development of Imaginal Dialogues*, 76.

BIBLIOGRAPHY

Bennett, J. G. "Arctic Origin of the Indo-European Culture." *Systematics* 1, no. 3 (1963): 203–32.

———. *Deeper Man*. London: Turnstone, 1978.

———. *The Dramatic Universe*, vol. 2, *The Foundations of Moral Philosophy*. London: Hodder and Stoughton, 1961.

———. *The Dramatic Universe*, vol. 4, *History*. London: Hodder and Stoughton, 1966.

———. *Energies: Material, Vital, Cosmic*. London: Coombe Springs Press, 1964.

———. *Gurdjieff, Making a New World*. London: Turnstone, 1964.

———. *Talks on Beelzebub's Tales*. London: Coombe Springs Press, 1975.

Blake, Anthony. *A Gymnasium of Belief in Higher Intelligence*. Charlestown, W. Va.: DUVersity Publications, 2010.

Crowhurst, Howard. *Megalithes*. Plouharnel, Brittany: HCom Editions, 2007.

Crowhurst, Howard et al. "Le Sire Mégalithique du Manio à Carnac." Plouharnel, Brittany: ACEM, 2009.

de Santillana, Giorgio. *The Origins of Scientific Thought*. New York: Mentor, 1966.

de Santillana, Giorgio, and Hertha von Dechend. *Hamlet's Mill: An Essay Investigating the Origins of Human Knowledge and Its Transmission through Myth*. Boston: David R. Godine, 1977.

Douglas, Mary. *Thinking in Circles*. New Haven, Ct.: Yale University Press, 2007.

Goethe. *Der Zauberlehrling*. n.p., 1797.

Guillaumont et al., trans., *The Gospel According to Thomas*. New York, Harper and Row, 1959.

Gurdjieff, G. I. *All and Everything: Beelezebub's Tales to His Grandson*. London: Routledge & Kegan Paul, 1973.

———. *The Herald of Coming Good.* New York: Weiser, 1933.

Haritayana, trans., and Swami Ramananda Saraswathi. *Tripura Rahasya or The Mysteries beyond the Trinity,* 4th edition. Tiruvannamalai: Sri Ramanashram, 1980.

Hawkins, Gerald. *Stonehenge Decoded.* London: Souvenir, 1966.

Heath, Richard. *Matrix of Creation.* Rochester, Vt.: Inner Traditions, 2002.

———. *Sacred Number and the Origin of Civilization.* Rochester, Vt.: Inner Traditions, 2007.

Heath, Robin. *Cracking the Cosmic Code.* Cardigan, Wales: Bluestone, 2008.

———. "The Discovery of a Soli-Lunar Calendar Device within an Astronomical Ritual Complex at Le Manio, Morbihan, Brittany," June 2009, available at www.skyandlandscape.com. Accessed on September 30, 2010.

———. *Sun, Moon and Stonehenge: Proof of High Culture in Megalithic Britain.* Cardigan, Wales: Bluestone, 1998.

Heath, Robin, and John Michell. *The Measure of Albion.* Cardigan, Wales: Bluestone, 2004.

Hoyle, Fred. *On Stonehenge.* San Francisco: W. H. Freeman, 1977.

Marshack, Alexander. *The Roots of Civilisation: Cognitive Beginnings of Man's First Art Symbol and Notation.* London: Weidenfeld and Nicolson, 1972.

McClain, Ernest. *The Myth of Invariance: The Origin of the Gods, Mathematics and Music from the Reg Veda to Plato.* New York: N. Hayes, 1976.

———. "The Tonal Calendar" in *The Myth of Invariance.* Boulder, Colo.: Shambhala, 1977.

Michell, John. *The Dimensions of Paradise.* Rochester, Vt.: Inner Traditions, 2007.

Ouspensky, P. D. *In Search of the Miraculous.* London: Routledge & Kegan Paul, 1980.

Rowlands, Peter. *Zero to Infinity.* London, World Scientific Publishing, 2007.

Schwaller de Lubicz, R. A. *The Temple of Man.* Rochester, Vt.: Inner Traditions, 1998.

Sullivan, William. *The Secret of the Incas.* New York: Crown, 1996.

Thom, A., and A. S. Thom. *Megalithic Remains in Britain and Brittany.* Oxford, England: Clarendon Press, 1978.

Tilak, Bal Gagadhar. *The Arctic Home in the Vedas.* Poona: Tilak Bros, 1903.

———. *The Orion.* Bombay: R. A. Sangoon, 1893.

Watkins, Mary. *Invisible Guests: The Development of Imaginal Dialogues.* Boston: SIGO Press, 1990.

Yukteshwar, Shri. *The Holy Science.* Los Angeles: Self-Realisation Fellowship, 1977.

INDEX